Geotechnics for Catastrophic Flooding Events

Geotechnics for Catastrophic Flooding Events

Editor

Sidhrath Saini

Geotechnics for Catastrophic Flooding Events

Edited by **Sidhrath Saini**

Printed in 2017

ISBN: 978-1-68117-012-1

Library of Congress Control Number: 2015931527

© 2016 by
SCITUS Academics LLC,
616, Corporate Way, Suite 2, 4766,
Valley Cottage, NY 10989

www.scitusacademics.com

Contents

Preface

Flood management planning and design of flood control facilities has evolved in recent years. Today, management of floods include planning and construction of facilities such as dams and detention facilities, coordinated operations of the reservoirs with flood control reservations, improvement of flood channels and levees, watershed management and proper land use planning.

Flood control refers to all methods used to reduce or prevent the detrimental effects of flood waters.

Floods are caused by many factors: heavy rainfall, highly accelerated snowmelt, severe winds over water, unusual high tides, tsunamis, or failure of dams, levees, retention ponds, or other structures that retained the water. Flooding can be exacerbated by increased amounts of impervious surface or by other natural hazards such as wildfires, which reduce the supply of vegetation that can absorb rainfall.

Geotechniques is the application of scientific and engineering principles to the solution of civil engineering and other problems created by the nature and constitution of the Earth's crust.

Geotechnics for Catastrophic Flooding Events is a reference book for researchers, academics, industry practitioners and other professional involved in earthquake geotechnical engineering, foundation engineering, and earthquake engineering and structural dynamics.

Editor

Ground Water Chemistry Changes before Major Earthquakes and Possible Effects on Animals

Rachel A. Grant, [1] Tim Halliday,[2] Werner P. Balderer,[3] Fanny Leuenberger,[4] Michelle Newcomer,[5] Gary Cyr,[6] and Friedemann T. Freund[5,6,7]

[1] Department of Life Sciences, the Open University, Milton Keynes, MK7 6AA, UK

[2] 21 Farndon Rd, Oxford OX2 6RT, UK

[3] Swiss Geotechnical Commission, Department of Earth Sciences, ETH Zurich, NO FO 35, 8092 Zurich, Switzerland

[4] Department of Earth Sciences, Geological Institute, ETH Zurich, NO G39.1, 8092 Zurich, Switzerland

[5] Ames Research Center, National Aeronautics and Space Administration (NASA), Earth Science Div. Code SGE, Moffett Field, CA 94035, USA

[6] Department of Physics, San Jose State University, San Jose, CA 95192, USA

[7] Carl Sagan Center, SETI Institute, 189 Bernardo Ave., Mountain View, CA 94043, USA

ABSTRACT

Prior to major earthquakes many changes in the environment have been documented. Though often subtle and fleeting, these changes are noticeable at the land surface, in water, in the air, and in the ionosphere. Key to understanding these diverse pre-earthquake phenomena has been the discovery that, when tectonic stresses build up in the Earth's crust, highly mobile electronic charge carriers are activated. These charge carriers are defect electrons on the oxygen anion sub lattice of silicate minerals, known as positive holes, chemically equivalent to O^- in a matrix of O^{2-}. They are remarkable inasmuch as they can flow out of the stressed rock volume and spread into the surrounding unstressed rocks. Travelling fast and far the positive holes cause a range of follow-on reactions when they arrive at the Earth's surface, where they cause air ionization, injecting massive amounts of primarily positive air ions into the lower atmosphere. When they arrive at the rock-water interface, they act as $\bullet O$ radicals, oxidizing water to hydrogen peroxide. Other reactions at the rock-water interface include the oxidation or partial oxidation of dissolved organic compounds, leading to changes of their fluorescence spectra. Some compounds thus formed may be irritants or toxins to certain species of animals. Common toads, *Bufo bufo*, were observed to exhibit a highly unusual behavior prior to a M6.3 earthquake that hit L'Aquila, Italy, on April 06, 2009: a few days before the seismic event the toads suddenly disappeared from their breeding site in a small lake about 75 km from the epicenter and did not return until after the aftershock series. In this paper we discuss potential changes in groundwater chemistry prior to seismic events and their possible effects on animals.

INTRODUCTION

Earthquakes are the most feared among all natural disasters because they seem to strike suddenly, without any forewarning. However, there have been innumerable reports of non-seismic pre-earthquake signals hours, days, and sometimes even weeks before major seismic events. These signals are often fleeting and subtle, seemingly "unreliable", but occasionally distinct and strong. Many of these pre-earthquake phenomena can reportedly be perceived by animals, eliciting unusual behavior. Reports relating unusual animal behavior to imminent earthquakes date back to antiquity as recounted in Tributsch's classic book "When the Snakes Awake" [1]. Recent reports of a wide range of physical pre-earthquake phenomena draw on modern science and technology involving ground-based and satellite-based observations. Here is a partial list:

- Low to ultralow frequency electromagnetic emissions from the ground,
- Luminous phenomena, often called earthquake lights, prior to many seismic events,
- Enhanced infrared emission from the epicentral region as seen in satellite images,
- Changes in the atmosphere near the ground and at altitudes up to about 12,000 m,

Perturbations in the ionosphere 100–600 km above the Earth's surface,

Changes in the ocean water and ground or spring water chemistry, *etc.*

Until recently the field of non-seismic pre-earthquake signals was in a general state of confusion. Nobody seemed to be able to identify a physical process, or a sequence of processes, capable of explaining the diversity of the reported pre-earthquake signals or how they may be linked to each other and/or traced back to a physical cause. This lack of understanding has been largely overcome by the discovery of a previously unknown form of electrification when rocks are subjected to mechanical stress [2–4].

LABORATORY AND FIELD OBSERVATIONS

Positive Whole Charge Carriers in Rocks

Rocks are generally thought of as good insulators. However, essentially all rocks in Earth's crust contain fundamental and seemingly ubiquitous types of defects, which had not been previously recognized. These defects are dormant and electrically inactive. In silicate minerals they consist of proxy links, e.g., of sites in the structures of common silicate minerals where normal $Si/^O\backslash Si$ bonds between $[SiO_4]$ structural units are replaced by $Si/^{OO}\backslash Si$. These are peroxy links and their characteristic feature is that the two oxygen's, which form the peroxy bond, have changed their valence state from the usual O^{2-} to the unusual O^-.

In the language of semiconductor physics an O^- in a matrix of O^{2-} represents a defect electron or hole, conventionally written as h•, also known as positive hole [5]. In the peroxy bond two positive holes are tightly bound, hence electrically inactive. However, when a rock is subjected to mechanical stress, dislocations move through the mineral grains. These moving dislocations cause the peroxy bonds to momentarily break. Scheme 1 depicts how this bond breakage proceeds in two steps, forming at first a short-lived transient state where the two positive holes decouple, followed by an electron transfer from an O^{2-} in the neighborhood. Once transferred, this electron becomes trapped in the broken proxy bond, while the O^{2-} that has donated the electron changes into an O^-. It becomes a positive hole h•, an electronic charge carrier.

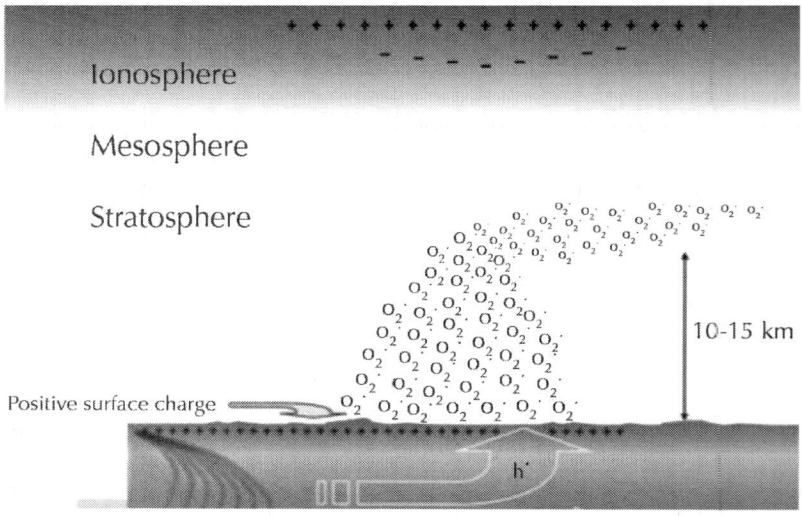

Scheme 1: Peroxy bond breakage and generation of a positive hole, h•, which become a mobile electronic charge carrier.

The important point to note is that, in the reaction sequence shown in Scheme 1, two electronic charge carriers are generated: a hole h• and an electron e'. They are both long-lived. As electronic states associated with defect electrons in the O^{2-} sub lattice, the h• can move away from the sites, where they were formed. They can propagate not only through the given mineral grain but also jump from grain to grain, spreading from the stressed rock into the unstressed rock [4].

This process can be demonstrated in the laboratory by stressing a rock at one end as illustrated in Figure 1a and following the h• as they spread into the unstressed rock. The stressed sub volume, from where the h• flow out, becomes negatively charged. The unstressed rock becomes positively charged. This creates an electrical potential.

$$O_3Si\!\!\diagup^{OO}\!\!\diagdown SiO_3 + O^{2-} \rightleftharpoons O_3Si\!\!\diagup^{O:}\!\!{}_O\!\!\diagup SiO_3 + O^{2-} \rightleftharpoons O_3Si\!\!\diagup^{O}\!\!{}_O\!\!\diagup SiO_3 \quad O•$$

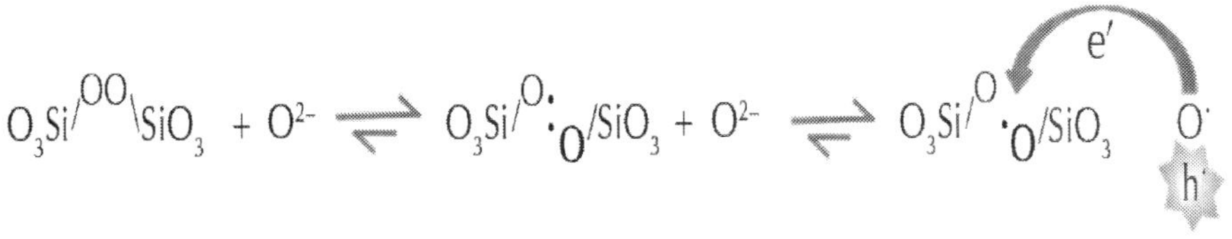

Figure 1: (a) The stressed rock volume turns into the source of charge carriers. The positive holes h• flow out into the unstressed rock, creating an electrical potential like in a battery. (b) The battery circuit can be closed by running a wire from the..

In order for the e' to also flow out, a separate pathway has to be provided. The situation resembles that in an electrochemical battery, where positively charged cations flow out of the anode into and through the electrolyte, while the electrons available in the anode have to wait until they are offered a metal wire to follow suit. In the case of the "rock battery" the positive charges are h•. Figure 1b shows that, if a wire is attached to the pistons, which are in electrical contact with the stressed rock volume, and to a Cu electrode at the unstressed end of the rock, the battery circuit is closed allowing a current to flow.

The h• are highly mobile. Once created as illustrated in Scheme 1 they are thought to propagate by way of a phonon-mediated electron transfer mechanism [6]. The experimentally determined maximum speed with which h• charge clouds are able to propagate through solid rock is on the order of 200 m/s, consistent with such a phonon-mediated electron transfer mechanism [4].

Prior to earthquakes tectonic stresses in the Earth's crust increase, eventually reaching the breaking point of the rocks. During the stress build-up h• charge carriers become activated in increasing numbers. The stressed rock volume thus turns into a battery, from where an h• current can flow out. This is the basic process that provides a comprehensive understanding of a wide range of pre-earthquake signals [2].

The h• streaming out into the surrounding unstressed rocks can travel fast and far, meters in laboratory experiments and probably tens of kilometers in the field. If the battery circuit closes, h• currents can flow continuously, in a quasi-dc mode. Under certain conditions they tend to fluctuate, providing a physically reasonable mechanism to account for the widely reported pre-earthquake ultralow frequency electromagnetic emissions [7–9].

Positive holes flow not only through solid rocks but also through sand and soil. When they reach the Earth surface, the h• can pairwise recombine to form peroxy bonds again. This is an exothermal reaction. It leads to vibrationally highly excited states, which de-excite by emitting infrared photons [10]. The infrared emission thus created may be responsible for the widely reported pre-earthquake "thermal anomalies" captured in night-time infrared satellite images of the areas around future epicenters [11–13]. In addition the arrival of h• charge carriers at the surface leads to the buildup of surface/subsurface charges, which generate microscopic electric fields, strong enough to ionize the air and inject massive amounts of airborne ions at the ground-to-air interface [14]. The surface charges can cause corona discharges accompanied by visible light [4, 15] and by noise in the radiofrequency range, which can interfere with telecommunications [16].

Air laden with positive ions leads to condensation of water droplets, causing haze and clouds. Expanding upward, the ionized air will carry along the Earth's ground potential to stratospheric heights. The resulting changes in the vertical electric field are expected to affect the ionosphere and may be the cause of ionosphere perturbations [17–19] and changes in the transmission of radio waves [20–22], which reportedly precede major earthquakes by a few days.

Positive airborne ions cause the blood serotonin level to increase, both in humans and animals [23]. Thus, changes in the serotonin levels as result of a pre-earthquake injection of positive airborne ions into the Earth's near-surface atmosphere may be instrumental in causing the widely reported anomalous behavior of land animals prior to major earthquakes [14, 24, 25], in causing an increase in migraine incidences in humans [26] and the general response of humans to air masses laden with positive ions [27, 28].

In this paper we consider what happens when h• charge carriers flow into water, and what type of chemical reactions occur at the rock-water interface that change the water chemistry and may affect animal behavior.

Oxidation of Water to Hydrogen Peroxide

Figures 2a/b depict that, while an h• moves in one direction, electrons e' hop in the opposite direction through a succession of O^{2-}, which turn momentarily into O^-. Figures 2a/b illustrate five electron transfers from right to left, while the h• moves five positions from left to right.

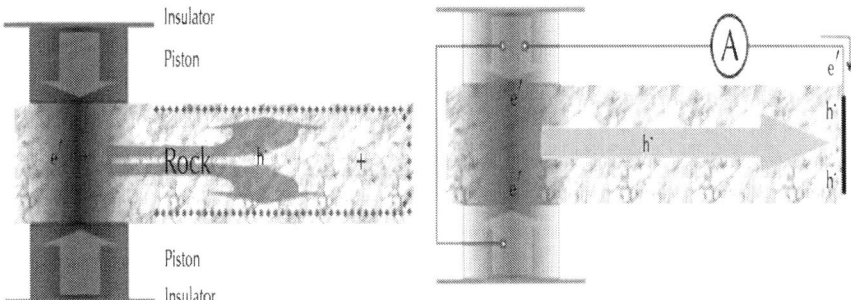

Figure 2: Schematic representation of the flow of a positive hole, e.g., a defect electron on the O^{2-} sublattice through a sequence of electron transfer steps from left to right. (a) at the rock-water interface the positive hole appears as an O^-

Physically an h• is an electronic charge carrier but chemically it is an O^-, an open shell configuration with 7 electrons in its outer shell. Such an O^- acts as a highly oxidizing •O radical, in biochemistry generically known as "reactive oxygen species", ROS.

Upon arriving at the rock-water interface, as depicted in Figure 2a, the h• acts as O^-, capable of oxidizing an H_2O molecule. As the

O^-subtracts an H from H_2O, it changes into an OH^-, which remains imbedded in the rock surface. At the same time the H_2O molecule turns into an •OH radical as depicted by Figure 2b. Two •OH radicals combine to form hydrogen peroxide:

$$O^-_{surface} + H_2O_{solution} \overline{Y}OH^-_{surface} + \bullet OH_{solution} \qquad (1a)$$

$$\bullet OH_{solution} + \bullet OH_{solution} \overline{Y}O_2H_{2(solution)} \qquad (1b)$$

Open Circuit versus Closed Circuit

As $h^\bullet$ flow from one point to another they create a potential difference. The closest analog is the outflow of captions from the anode of an electrochemical battery creating a "battery voltage". When the battery circuit is open, as depicted in Figure 1a, the potential counteracts the charge outflow and soon brings it to a halt. Hence, the number of charge carriers that can flow out of the anodic volume is limited. In order to establish a continuous outflow current, it is necessary to close the circuit, as depicted in Figure 1b, allowing electrons to also flow out. This analogy also applies to the rock battery.

Of course, in the Earth's crust, there are no metal wires connecting different volumes of rock that are subjected to different levels of tectonic stress. However, Nature provides other alternatives to close the "rock battery" circuit [2]. One possibility is to establish an electrolytically conductive pathway. Figure 3a illustrates a laboratory set-up where the $h^\bullet$ are allowed to flow from the stressed rock volume into a water reservoir attached to the unstressed end of the rock sample. As the $h^\bullet$ enter the water, they react to form H_2O_2 at the rock-water interface as described by Equations 1a,b, but the current continues through the bulk of the water, presumably by way of H_3O^+ [29]. When the H_3O^+ reach the Cu electrode, they are expected to react with the electrons that have traveled through the metal wire from the stressed rock volume to the Cu electrode:

$$2e' + 2H_3O + \overline{Y} 2H_2O + H_2 \qquad (2)$$

(a)

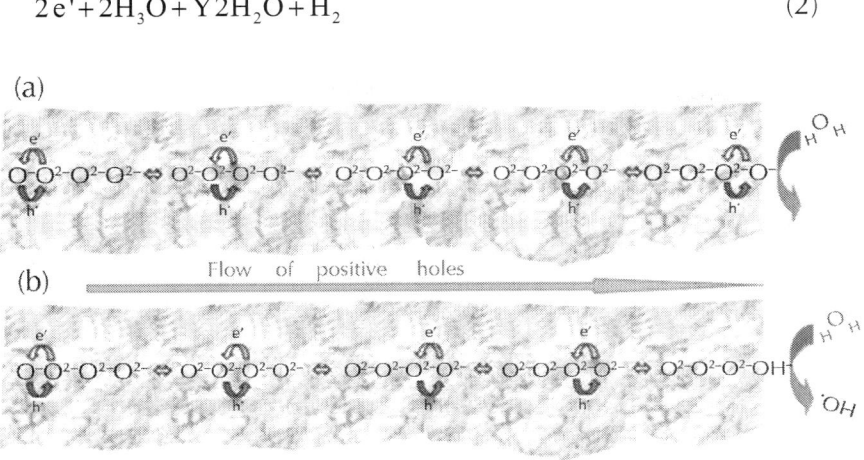

(b)

Flow of positive holes

Figure 3: (a) Basic laboratory set-up to demonstrate that stress-activated h• currents can flow into water and close the battery circuit. (b) Demonstration of a rock battery where the circuit closure is achieved through the electrolytical conductivity of water.

Equation 2 predicts that closure of the rock battery circuit is possible by electrolyzing water to H_2O_2 plus H_2 [29]. If this is true, we can get rid of the metal wire plus Cu electrode altogether and close the rock battery circuit by connecting the water reservoir at the unstressed end of the rock with a water reservoir that is in contact with the stressed rock volume.

Figure 3b illustrates the experimental set-up used to demonstrate closure of the rock battery circuit through an entirely electrolytical pathway. Formation of H_2O_2 in the water reservoir at the unstressed end of the rock was used to verify that the circuit closure had indeed taken place [29].

Electrocorrosion of Rocks

According to Equation 1a the surface of a rock, through which h•charge carriers flow into water, must become hydroxylase. This is equivalent to saying that the rock undergoes accelerated weathering or some form of electro-corrosion. Expected consequences of such a process are that, while h• flow into the water, the pH values become more acidic and more captions will be released from the rock surface than during normal rock-water interactions.

There have been reports in the literature that, prior to earthquakes, the concentration of dissolved captions in spring and ground water often increases, for instance at ground water measuring stations along the North Anatolian Fault in Turkey [30]. Moderate seismic activity in the otherwise stable Northwestern part of Spain was preceded by several weeks by distinct changes in the water chemistry including increases in electrical conductivity due to dissolved captions and anions [31]. Likewise the pH values in deep boreholes in Kamchatka, Russia, reportedly indicated increased acidity of deep well waters prior to major earthquakes in the region [13, 32].

The electrocorrosion of rocks can be demonstrated in the laboratory. Figure 4 depicts a slab of gabbro, $30 \times 60 \times 9$ cm^3, with identical water reservoirs attached to both ends. Both reservoirs contain a 20×6 cm^2 Cu electrode separated from the rock surface by 1 cm of water. Applying a constant load of about 10 tons to the center of the gabbro slab through a pair of stainless steel pistons, $5 \times 5 \times 30$ cm^3, activates h• charge carriers and causes them to flow out as indicated by gray arrows. Figure 4 shows circuit closure for the water reservoir on the right but not for the one on the left.

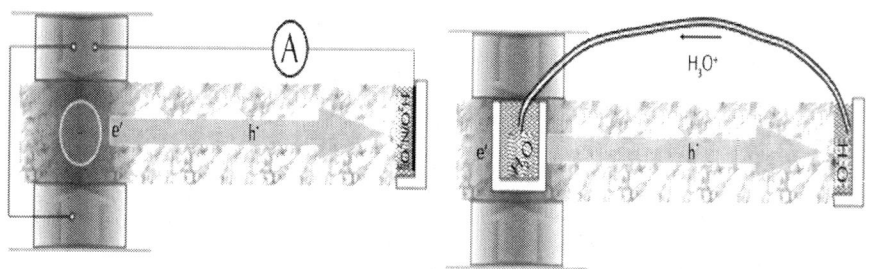

Figure 4: Experimental set-up to demonstrate the accelerated dissolution (electrocorrosion) of the rock surface while h• charge carriers flow into the water reservoir for which circuit closure has been established.

Figures 5a/b shows the differences in the concentrations of dissolved K^+, Ca^{2+} and Mg^{2+} measured between the two water baths as a function of time, while the load on the rock was kept constant. Starting in the second week the dissolved cation contents increased in the water bath for which circuit closure had been established, indicating accelerated dissolution of the rock surface.

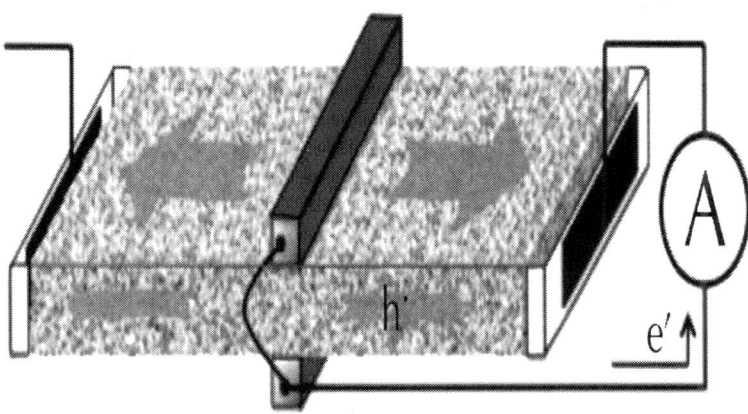

Figure 5: (a/b) Excess cation concentrations measured in the water reservoir attached to the slab of gabbro into which h• were allowed to flow continuously over a period of 10 weeks.

Other Oxidation Reactions at the Rock-Water Interface

If h• charge carriers have the capability to oxidize H_2O to H_2O_2 at the rock-water interface, there are reasons to expect that they will also have the capability to cause other oxidation reactions. Of special interest are oxidation reactions involving organic compounds of biogenic origin dissolved in spring and ground water.

Groundwater samples were collected in July 1999 within the zone of influence of the North Anatolian Fault Zone (NAF). After the highly destructive M7.6 Izmit earthquake of August 17, 1999 further samples were taken in order to study possible effects of this earthquake on the isotopic and chemical compositions of thermal and mineral waters in the surrounding areas [33]. Significant changes were also observed in the intensities of the fluorescence spectra of thermal and mineral waters collected prior to the Izmit earthquake in the nearby areas of

Kuzuluk, Bursa, and Yalova/Gemlik [34]. Figure 6 shows fluorescence spectra of water samples from a location near Bursa, Turkey. Sample #16, collected July 10, 1999, 7 weeks before the earthquake, exhibits a slightly elevated 340 nm intensity.

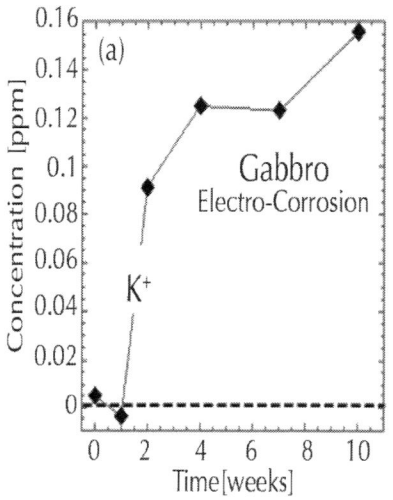 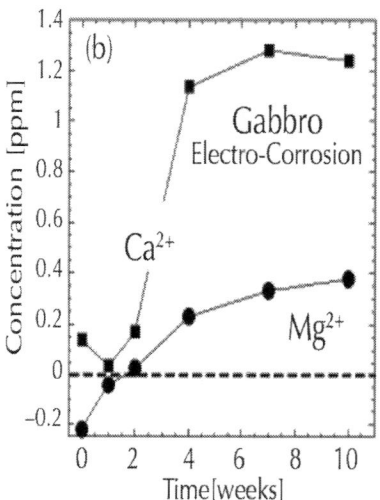

Figure 6: Fluorescence spectra of water samples of the Cekirge (Bursa) Vakif Bahçe Spring, Turkey (sampling 1999 to 2003).

In sample #33, collected October 6, 1999, during intense aftershock activity about 7 weeks after the Izmit earthquake, the 340 nm peak intensity is significantly higher. One year later, in sample #47, the fluorescence intensity at 340 nm had decreased, and remained low over the course of the next 3 years, though sample #90, collected Aug. 2003, exhibited a higher fluorescence intensity around 280 nm.

Another example comes from the Lago di Garda, Italy, where an M5.3 earthquake occurred on Nov. 24, 2004, at a focal depth of 25 km, 8 km from the village of Salò [35]. Figure 7 shows the fluorescence spectrum #1 of a reference water sample from the Tavina Mineral Spring at Salò and the spectra of several water samples collected about 17 days before the earthquake, #2, #3, and #4 from three open outlets of the Tavina spring, #5 from a fountain on the road to Gardone, and #6 from an adjacent creek.

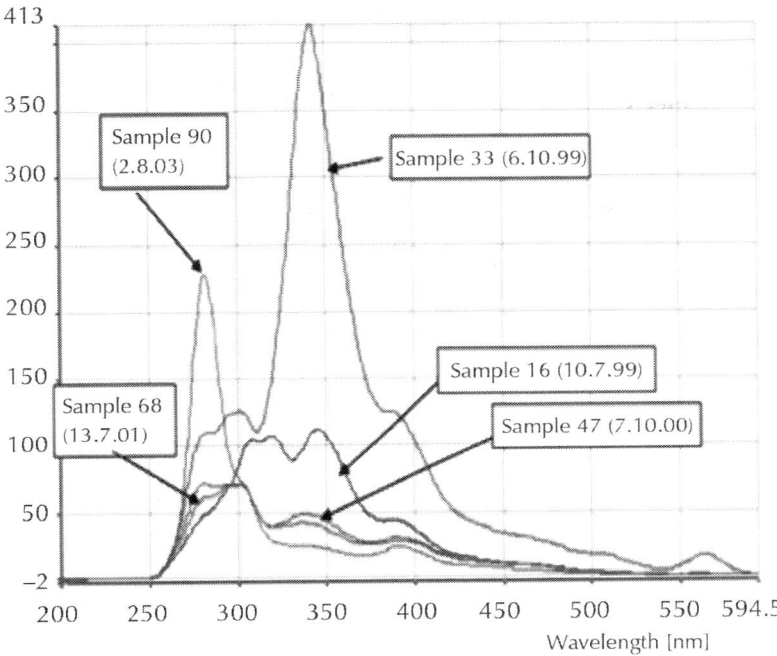

Figure 7: Water samples collected on the shores of Lago di Garda, Italy, prior to and after a modest M5.3 earthquake, 25 km focal depth, 8 km from Salò.

The observed changes in the florescence indicate that existing dissolved organic compounds spectra in the groundwater and spring water become partially oxidized. A well-known example of this type of reaction is the oxidation of terephthalate, an aromatic dicarboxylic acid, used as an assay for ROS, reactive oxygen species [36]. As Figure 8 shows the reaction with an •O radical leads to the addition of an O to the aromatic ring, changing the fluorescence of the molecule.

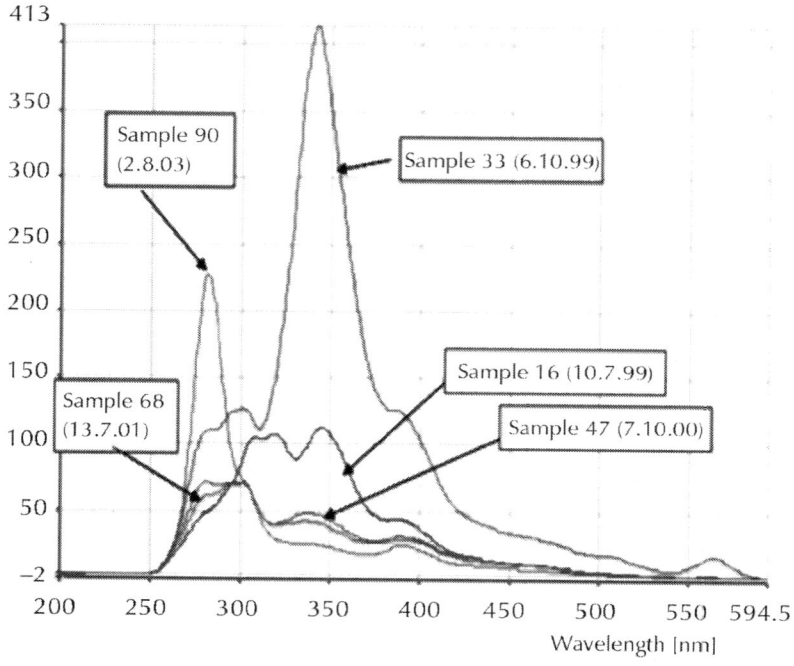

Figure 8: Terephthalate assay: partial oxidation of the aromatic ring through the reaction with an •O radical, leading to a diagnostic change in the fluorescence spectrum [36].

Which specific organic compounds were affected in the groundwater and spring water samples depicted in Figures 5 and 6is not known at this time However, the observations points to oxidative reactions by h• charge carriers that become activated in the stressed rocks within the hypocentral region of an impending earthquake or during the aftershock series. These h• charge carriers will flow down stress gradients and become available over a wide area, tens of kilometers in diameter, surrounding the sites of maximum seismic activity. While propagating through the near-surface soil, the h• will enter the water table from below. They are expected to cause a wide range of oxidation reactions. If these reactions involve organic matter dissolved in the groundwater or adsorbed to soil particles, the products may include partially oxidized ketones and carboxylic acids, which would remain in solution, all the way to carbon monoxide, CO, which can be emitted as a gas from the soil.

Massive amounts of CO were released from the ground 4–6 days before the M7.6 Gujarat earthquake of January 26, 2001 in Northwest India, enough to be detected in spectrally resolved daytime and night-time images of the MOPITT sensor onboard the NASA TERRA satellite [37]. Since the CO concentration was highest at ground level, reaching an average of 0.25 ppmv (volume parts per million) between ground and about 6000 m altitude, the CO was clearly associated with an emanation from the soil.

Unusual Animal Behavior before Earthquakes

Unusual animal behavior before major earthquakes has been reported through the ages [1,38]. Apparently, during the build-up of stresses deep in the Earth crust to dangerously high levels, many animals are able to perceive cues from the environment, which cause them to react abnormally. Animals both on land and in water are reportedly affected. Evidence for unusual animal behavior has been widely reported in the past as a warning sign of impending major earthquake activity, though such reports have widely been called anecdotal [24,39]. Since earthquakes are rare events, reports of unusual animal behavior are—by their very nature—in almost all cases anecdotal [24,39].

Attempts to validate the unusual animal behavior in laboratory experiments have been largely inconclusive [25, 40]. However, as pointed out in a recent review of the 1975 Haicheng earthquake in China [41]: "Among the animals, the most difficult to ignore are the snakes coming out hibernation dens when the average temperature was much below freezing. There were nearly 100 snake sightings within one month prior to the earthquake [42] such suicidal behavior is extremely difficult to explain".

A recent example of unusual animal behavior has been given by Grant and Halliday, who reported that the activity of common toads at a breeding site in a small lake about 75 km north of L'Aquila, Italy, declined dramatically five days before the M6.3 L'Aquila earthquake on April 6, 2009 [43]. The apparent abandonment of the breeding site and interruption of spawning is highly unusual in these amphibians, which exhibit an explosive breeding behavior. Once the breeding activity has commenced, the toads normally do not leave the site for 3–7 weeks until spawning is completed [44]. Although the breeding site was some distance from the epicentre, it has been shown that there was a major extension of earthquake related phenomena to the north of the epicentre, including earthquake lights and electric anomalies [45]. A similar effect seemed to have occurred before the great 1873 earthquake in central Italy as reported by Serpieri, who observed unusual snake behavior in his laboratory upon application of electric currents [46].

Figure 9(top) shows for the period of March 26 to April 17, 2009 the number of male toads observed per day. Figure 9(bottom) shows the L'Aquila seismic sequence, which includes the M6.3 main shock at 3:32 am on April 06 [43]. During 5 days leading up to the main shock the number of male toads observed dropped to near-zero. Their number recovered slightly after the main shock and dropped again, though not to near-zero, prior to the major aftershock of April 13.

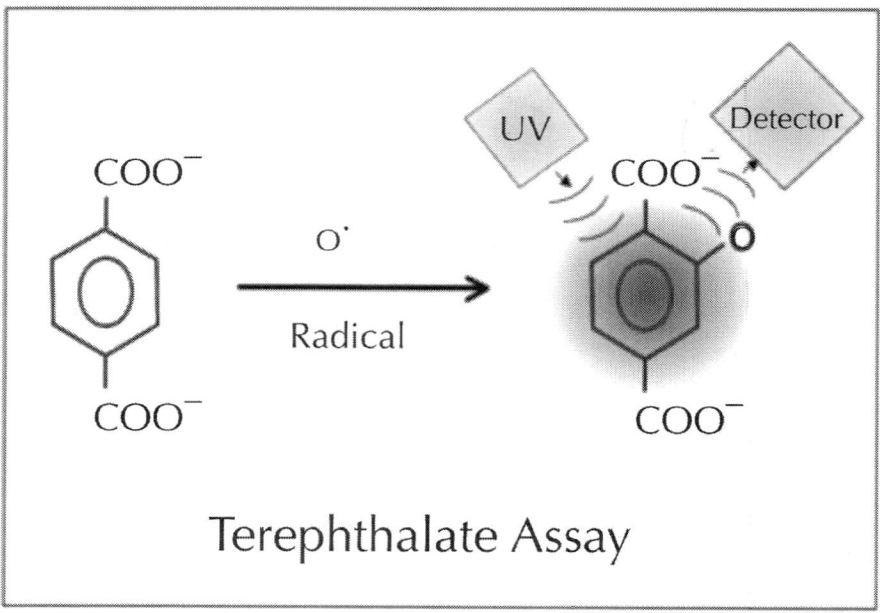

Figure 9: (top) Number of male toads observed at the Lake Site; (bottom) Seismic activity associated with the M6.3 L'Aquila earthquake of April 06, 2009 including foreshock, main shock and aftershock activity.

Figure 10 shows the rainfall recorded both locally at the site and at a weather station about 15 km away from the lake. Since the period of highest rainfall between March 30 and April 04 coincides with the apparent disappearance of the toads from their breeding site, lack of precipitation can definitely not be the cause of the unusual toad behavior.

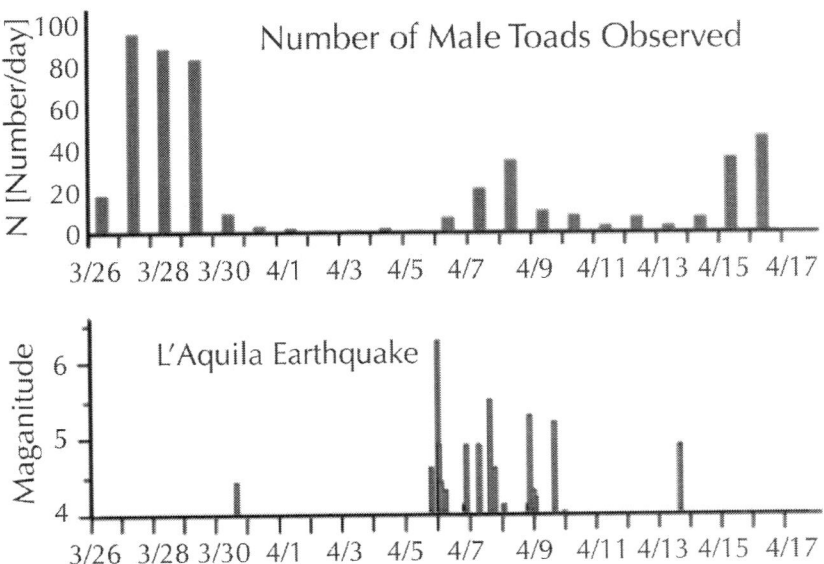

Figure 10: Rainfall records for the lake site and at a weather station at a distance of 15 km.

The rainfall measurements at the breeding site and contemporary field observations indicate that the ground was highly saturated in the week prior to the earthquake. Under normal circumstances such conditions would be advantageous to toads, which are prone to desiccation because of their permeable skins and always prefer humid or damp environments. The disappearance of the toads from the breeding site is all the more unusual, as they would be expected to be more active during rainy periods [47]. The time period during which the toads left the lake coincided with other suspected pre-earthquake indicators such as disturbances in the ionosphere. These disturbances consist of changes in the transmission characteristics of radiowaves along direct great circle paths between two stations. The changes arise between sunset and sunrise, e.g., during the dark hours, when the ionosphere above the region of interest, in this case central Italy, moves out of and back into the influence of the ionizing solar radiation, respectively. Processes that occur at the Earth surface before major earthquakes have an influence on the concentration profile of free electrons in the ionospheric plasma [48,49]. Electrons contribute primarily to the mirror-like reflection of radiowaves, which allows them to be transmitted over long distances. Therefore a change in the vertical distribution of electrons in the ionosphere causes a distinct shift in the so-called terminator times, e.g., in the transmission characteristics around sunset and sunrise [50, 51].

Figure 11 shows an example of such ionospheric disturbances as they expressed themselves in the difference between the amplitudes (dAmplitudes) of radio signals emitted from two stations in Italy, ICV at 20.27 kHz in Sardinia and ITS at 45 kHz in Sicily, and received at the MOS station in Moscow, Russia. On their path to MOS in Moscow, Russia, the ICV and ITS signals pass over the ionospheric region that lies within the sphere of influence of the L'Aquila earthquake [20]. By referencing the signals along the path ICV-MOS and ITS-MOS to signals received in Moscow from radio stations in England and Iceland, it can be shown that the ionosphere above central Italy was perturbed several days before the L'Aquila event, giving rise to anomalous negative intensities (dAmplitude) as marked by dotted circles.

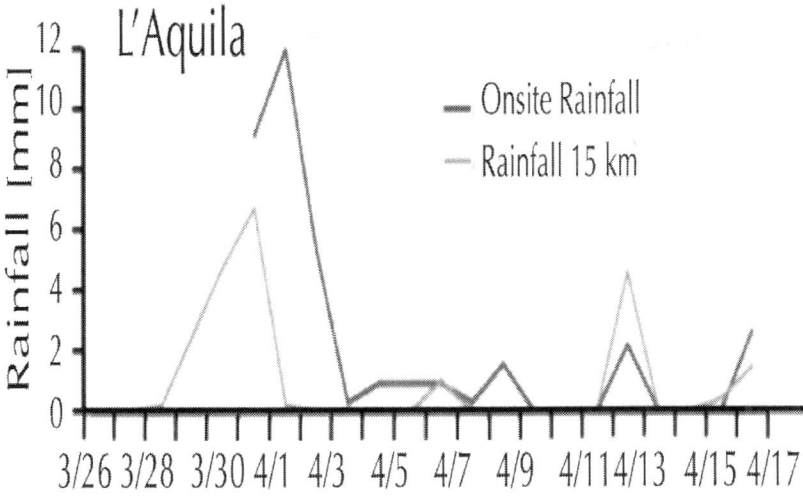

Figure 11: Evidence for ionospheric disturbances along the radio transmission paths between two emitter stations in Italy, ICV in Sardinia and ITS in Sicily, and a receiver station MOS in Moscow, Russia. The disturbances occur between sunset and sunrise and are

Unusual pre-earthquake conditions in the ionosphere above the region of L'Aquila as presented here [20] have been confirmed by other authors [52]. Electromagnetic signals ranging from ultralow to kHz and MHz frequencies have been recorded around the same time [53–55]. In addition there have been a number of reports of flashes of visible light and of "cold flames" breaking out of the ground at distances up to 50 km from the L'Aquila epicenter [45]. Such visible light phenomena have an electric origin and are known as "earthquake lights" [56, 57].

DISCUSSION

Unusual animal behavior before major earthquakes has been reported through the ages [1]. There are reports of anomalous behavior relating to fish, amphibians, and reptiles. Deep-sea fish rising to the surface have been observed on numerous occasions, as have fish jumping out of the water [58, 59]. The deep sea species *Trachipterus ishikawae*, a ribbon fish, and *Regalecus russelii,* the oarfish, are commonly seen before earthquakes [60].

Semi-aquatic animals have the capacity to leave the water, whereas fully aquatic species do not have this option. Crabs have been seen leaving the water in large numbers prior to earthquakes and unusual movements of frogs and toads have been noted as a possible pre-earthquake indicator [43, 58, 59], including mass migrations of toads prior to the M8.0 May 12, 2008 Sichuan Earthquake [61]. Similarly, prior to the M7.3 February 04, 1975 Haicheng earthquake in northern China, hundreds of snakes were observed coming out of hibernation in subzero temperatures [41], a behavior that is highly unusual and would generally be considered maladaptive in this ectothermic group of animals.

However, much unusual pre-seismic animal behavior probably goes unreported in the scientific literature for several reasons: (i) The topic has become unfashionable among biologists as *post hoc* evidence is largely viewed as selective recognition [60], (ii) isolated incidents of unusual behavior are difficult to link conclusively to earthquakes, and (iii) the unpredictable nature of earthquakes means that biologists are rarely at the scene to record at the time of seismic events. An exception to this last rule is provided by the fortuitous observation of unusual behavior of toads in the lake near L'Aquila, Italy, prior to the M6.3 April 6, 2008 earthquake, which was part of a systematic 4-year study of toad breeding behavior at this very location [43].

Earth is a highly dynamic planet. Plate tectonic movements continuously cause a waxing and waning of stresses in the Earth's crust [62]. The strain is usually accommodated by the slow deformation of rocks and/or soft sliding along faults. Occasionally, however, stresses in the crust build up to dangerously high levels leading to catastrophic failure of rocks and to earthquakes.

What had not been previously recognized is the fact that, during the build-up of stresses, electronic charge carriers are activated through the breakage of pre-existing, yet dormant peroxy defects in the matrix of rock-forming minerals [2]. As illustrated in Scheme 1 this process leads to positive hole charge carriers, h•, which have the unusual property that they are able to flow out of the stressed rock volume and to spread into the surrounding less stressed or unstressed rocks. The h• can travel fast and far, meters in laboratory experiments, presumably kilometers to tens of kilometers in the field. The h$^\bullet$ are not only highly mobile, but also have a dual nature. On one hand they are charge carriers, whose propagation through the Earth crust constitutes an electric current. On the other hand, because the h$^\bullet$ represent an O$^-$ state in a matrix of O^{2-}, they are chemically equivalent to $\bullet$O radicals. As such they are highly reactive and highly oxidizing, capable of executing a variety of reactions, when they arrive at some common discontinuities in the environment such as the rock-air interface and the rock-water interface.

At the rock-air interface the h$^\bullet$ build up surface/subsurface charges, which have long been suspected to become strong enough— at sufficiently high number densities—to "rip off" an electron from air molecules [63] and thus inject positive airborne ions into the atmosphere [14].

At the rock-water interface the h$^\bullet$ have been shown to be able to extract an H from H_2O molecules, thereby forming $\bullet$OH radicals, which in turn form H_2O_2 as depicted in Figures 2a/b [29].

These two basic processes, plus the fact that h$^\bullet$ flowing through the Earth crust constitute an electric current, allow us to look at the diversity of reported pre-earthquake phenomena in a new way and to see correlations between seemingly disjointed observations such as the anomalous behavior of toads at the peak of their breeding season in small lake near L'Aquila, Italy, and ionospheric disturbances above central Italy during the days before a M6.3 earthquake which inflicted heavy damage to the town of L'Aquila and loss of life.

We thus begin to understand why, during the propagation of h$^\bullet$charge carriers through the rocks, one should expect to see ultralow frequency electromagnetic emissions, such as reported in the literature [9, 64–66]. We also begin to understand why, upon arrival of large numbers of h$^\bullet$ at the rock-air interface over a region of tens of kilometers across, possibly even larger, the air becomes ionized and large amounts of airborne ions, primarily positive ions, are injected at the rock-air interface as experimentally demonstrated [14]. Figure 12 depicts schematically such a situation where tectonic stresses build up inside the Earth crust (lower left) leading to the activation of an ever increasing number of h$^\bullet$ charge carriers. The h$^\bullet$flow out of the stressed rock volume. They are believed to be able to spread tens of kilometers down stress gradients and to the Earth surface.

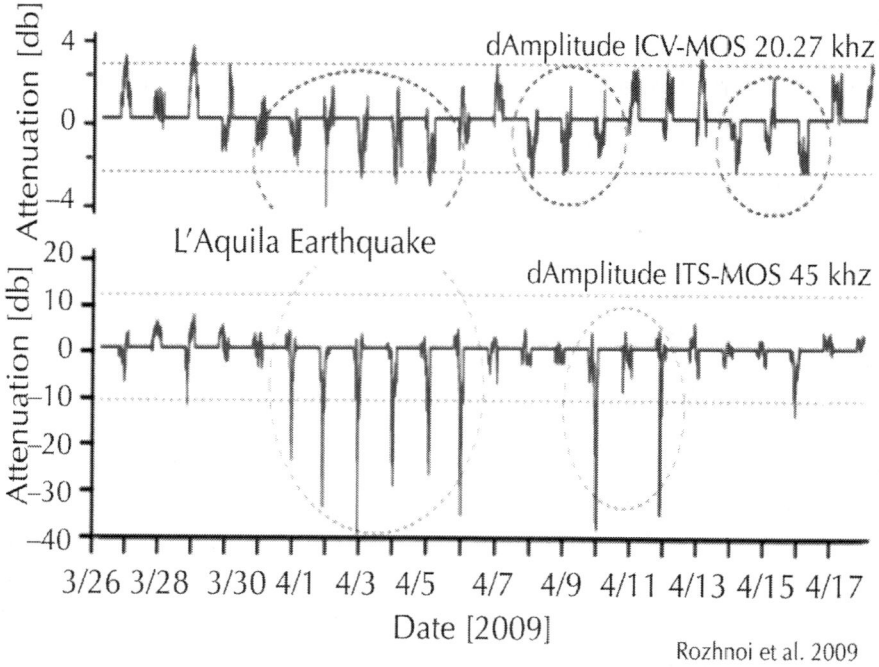

Rozhnoi et al. 2009

Figure 12: Concept drawing of the effect that an ion-laden air bubble, forming at ground level and rising through the atmosphere, will have on the ionosphere.

As the number density of h⋅ charge carriers at the surface increases, air molecules become field-ionized, probably O_2, generating large amounts of positive airborne ions at the rock-air interface. Prior to earthquakes episodes of air ionization have been observed [9]. Air laden with positive ions will expand upward, probably to stratospheric heights, dragging along the Earth's ground potential. The potential difference between the ground and the ionosphere is on the order of 250,000 V [67]. Therefore, as the ion-laden air bubble rises, it will alter the vertical electric field and cause the ionospheric plasma to respond accordingly. As indicated in Figure 12, electrons in the ionospheric plasma are predicted to be pulled downward. This will lead to an increase of the Total Electron Content (TEC) at the lower edge of the ionosphere [68–70] and to a change in the radio wave transmission characteristics as mentioned above. If waves of h⋅ charge carriers arrive at the Earth surface over a wide area surrounding a future epicenter and if the number of h$\cdot$ becomes so large in the days before a major earthquake as to ionize the air [14] and cause an ionospheric response, then we can expect that the same waves of h$\cdot$ charge carriers will enter into the water table and into bodies of water in the affected region. Crossing into the water the h$\cdot$ will oxidize H_2O to H_2O_2 as experimentally demonstrated [29], will oxidize or partially oxidize organic compounds dissolved in the water as suggested by the reported changes in the fluorescence spectra of ground and spring water samples from seismically active areas [34], and will acidify the water. Some partial oxidation products may have an O atom attached to an aromatic ring as in the case of terephthalate mentioned in the context of Figure 8 or may consist of ketones, carboxylic acids or CO. At the same time the pH is expected to drop as the water becomes acidified.

We will now discuss the potential effects of some of these changes on biological systems. Under non-extreme conditions, cells produce a variety of enzymes and non-enzyme antioxidants as a defense against oxidative stress, and there is a balance between oxidant and antioxidant processes. However, once the balance is upset, severe oxidative stress can lead to cell death by apoptosis and tissue necrosis [71, 72]. There are many studies, which demonstrate the time- and dose-dependent cytotoxic effects of H_2O_2 on lipids in the cell membrane [73], on cell proteins [74], and on DNA [75]. Hydroxyl radicals ($\bullet$OH) are highly reactive and very damaging, due to their capacity to react with any biological molecule, causing free radical chain reactions, the oxidation of membrane lipids and denaturing of proteins, thus inactivating important enzymes required for biological functioning [75]. Low pH has also been shown to detrimentally affect cells and whole animals for example, by reducing protein synthesis, which negatively affects growth and reproduction [76, 77]. Some of the oxidation products of dissolved organic matter are neurotoxins [78] and furthermore carbon monoxide can cause death by binding preferentially to hemoglobin instead of oxygen, forming carboxyhemaglobin, so that oxygen cannot be delivered to the tissues and organs of the body. High concentrations of cations may affect the water and sodium balance of aquatic animals.

Amphibians are particularly sensitive to changes in water chemistry as their skins are permeable to electrolytes [79]. Several anions and cations have been found to adversely affect the survival of amphibians at different life history stages [80]. The presence of certain metal ions increases the damaging effects of H_2O_2 [81] and exposure to H_2O_2 can lead to limb abnormalities in larval amphibians [82]. If CO is produced by the oxidation of organic matter in the soil or water column, it can lead to neuropsychological impairment, even at low levels [78]. Low pH in breeding ponds has been shown to cause both lethal and sublethal effects in a range of amphibian species [83] by inhibiting active uptake of Na^+ and increasing the loss of passive ions through amphibian skin [79]. Sodium ions (Na^+) are actively transported across amphibian skin from the environment (Cl^- follows passively) and the presence of high concentrations of H^+ K^+, Mg^{2+} and Ca^{2+} interferes with this process [84]. The inability of amphibians to maintain their sodium balance can quickly lead to death.

The combined evidence provided by the toad observation at the lake near L'Aquila and the ionospheric disturbance data as derived from radiosounding suggests that the toads were able to perceive in their environment some pre-seismic cues, which warned them of the impending earthquake. The toads may have responded to pre-seismic signals as a simple avoidance reaction to an adverse stimulus or as an evolved adaptation [25]. A possible reason for the toads' apparent movement to higher ground [43] might be found in an evolutionarily imprinted response to the danger of landslides and flooding. However, both of these explanations lack credibility, in particular the flooding argument because common toads are semi-aquatic during the breeding season and would be unlikely to leave flooded low lying land around the lake.

The analysis presented here approaches the coincidence of unusual toad behavior and ionospheric disturbances from a different perspective. Both phenomena are driven by a physical process in the Earth crust, in the future hypocentral volume, by the activation of h• charge carriers during the rapid increase in tectonic stress prior to the seismic event. As h• charge carriers spread out in ever larger numbers from the most severely stressed rock volume deep below, they cause different secondary processes at the land surface and at the rock-water interface. The process at the land surface consists of massive air ionization, which has a measurable effect on the ionosphere. The process at the rock-water interface consists of changes in the groundwater and presumably lake water chemistry, which seems to have provided the toads with an impetus to leave and seek refuge on higher ground. Thus, the toads did not react in response to the ionospheric disturbances *per* se but to cues that resulted from chemical reactions at the ground-water interface, which were driven by the arrival of stress-activated h• charge carriers from deep below. Their flight from adverse or toxic environmental conditions provides a more parsimonious explanation than the suggestion that animals might possess an evolved response to specific pre-earthquake conditions [25].

There is little doubt that anomalous animal behavior does occur prior to major earthquakes. Given the variety of physical and chemical processes documented to occur over the large earthquake preparation zone [85], it would in fact be surprising if animals were not affected. In this paper, we have suggested a possible common mechanism for diverse physical pre-earthquake processes and incidents of anomalous animal behavior. How this information may be used for forecasting earthquake risk will be the subject of future research.

ACKNOWLEDGMENTS

This work received partial support from NASA through Exobiology and Earth Surface and Interior grants to FTF and technical help from John Segre to, Jerry Wang and Lynn Hofland at the NASA Ames Research Center. RAG acknowledges funding from the Societas Europaea Herpetologica for her fieldwork in Italy, help from J. Taylor and S. Taylor during her toad observation project and from Alexander Rozhnoi, who patiently explained radio sounding techniques to detect ionospheric perturbations.

REFERENCES

1. Tributsch H. When the Snakes Awake: Animals and Earthquake Prediction. MIT Press; Cambridge, MA, USA: 1984. p. 264.

2. Freund FT. Toward a unified solid state theory for pre-earthquake signals. Acta Geophys. 2010;58:719–766.

3. Freund FT, Takeuchi A, Lau BW. Electric currents streaming out of stressed igneous rocks—A step towards understanding pre-earthquake low frequency EM emissions. Phys. Chem. Earth.2006;31:389–396.

4.   Freund F. Charge generation and propagation in rocks. J. Geodyn. 2002;33:545–572.

5.   Griscom DL. Electron spin resonance. Glass Sci. Technol.1990;48:151–251.

6.   Shluger AL, Heifets EN, Gale JD, Catlow CRA. Theoretical simulation of localized holes in MgO. J. Phys. Condens. Matter.1992;4:5711–5722.

7.   Kushwah V, Singh B, Hayakawa M. Ultra low frequency (ULF) magnetic field anomalies observed at Agra and their relation to moderate seismic activity in Indian region. J. Atmos. Sol.-Terr. Phys. 2005;67:992–1001.

8.   Fraser-Smith AC. Ultralow-frequency magnetic fields preceding large earthquakes. EOS. 2008;89:211.

9.   Bleier T, Dunson C, Maniscalco M, Bryant N, Bambery R, Freund FT. Investigation of ULF magnetic pulsations, air conductivity changes, and infra red signatures associated with the 30 October 2007 Alum Rock M5.4 earthquake. Nat. Hazards Earth Syst. Sci. 2009;9:585–603.

10. Freund FT, Takeuchi A, Lau BWS, Al-Manaseer A, Fu CC, Bryant NA, Ouzounov D. Stimulated thermal IR emission from rocks: Assessing a stress indicator. eEarth. 2007;2:1–10.

11. Tramutoli V, Cuomob V, Filizzolab C, Pergolab N, Pietrapertosa C. Assessing the potential of thermal infrared satellite surveys for monitoring seismically active areas: The case of Kocaeli (Izmit) earthquake, August 17, 1999. Remote Sens. Environ.2005;96:409–426.

12. Ouzounov D, Freund FT. Mid-infrared emission prior to strong earthquakes analyzed by remote sensing data. Adv. Space Res.2004;33:268–273.

13. Tronin AA, Molchanov OA, Biagi PF. Thermal anomalies and well observations in Kamchatka. Int. J. Remote Sens.2004;25:2649–2655.

14. Freund FT, Kulahci I, Cyr G, Ling J, Winnick M, Tregloan-Reed J, Freund MM. Air ionization at rock surface and pre-earthquake signals. J. Atmos. Sol.-Terr. Phys. 2009;71:1824–1834.

15. Araiza-Quijano MR, Hernández-del-Valle G. Some observations of atmospheric luminosity as a possible earthquake precursor. Geofis. Int. 1996;35:403–408.

16. Kolvankar VG. Report BARC-2001/E/006: Earthquake sequence of 1991 from Valsad Region, Guajrat. Bhabha Atomic Research Centre, Seismology Div; Mumbai, India: 2001. p. 68. BARC-2001/E/006.

17. Pulinets S, Boyarchuk K. Ionospheric Precursors of Earthquakes. Springer; Heidelberg, Germany: 2004. p. 350.

18. Liu JY, Chen CH, Chen YI, Yen HY, Hattori K, Yumoto K. Seismo-geomagnetic anomalies and M5.0 earthquakes observed in Taiwan during 1988–2001. Phys. Chem. Earth. 2006;31:215–222.

19. Nemec F, Santolík O, Parrot M, Berthelier JJ. Spacecraft observations of electromagnetic perturbations connected with seismic activity. Geophys Res Lett. 2008;35:L05109. doi: 10.1029/2007GL032517

20. Rozhnoi AM, Solovieva OM, Schwingenschuh K, Boudjada M, Biagi PF, Maggipinto T, Castellana L, Ermini A, Hayakawa M. Anomalies in VLF radio signals prior the Abruzzo earthquake (M = 6.3) on 6 April 2009. Nat. Hazards Earth Syst. Sci. 2009;9:1727–1732.

21. Kasahara Y, Muto F, Horie T, Yoshida M, Hayakawa M, Ohta K, Rozhnoi A, Solovieva M, Molchanov OA. On the statistical correlation between the ionospheric perturbations as detected by subionospheric VLF/LF propagation anomalies and earthquakes. Nat. Hazards Earth Syst. Sci. 2008;8:653–656.

22. Hayakawa M. Electromagnetic phenomena associated with earthquakes: A frontier in terrestrial electromagnetic noise environment. Recent Res. Dev. Geophys. 2004;6:81–112.

23. Krueger AP, Reed EJ. Biological impact of small air ions.Science. 1976;193:1209–1213.

24. Logan JM. Animal behavior and earthquake prediction. Nature.1977;265:404–405.

25. Kirschvink JL. Earthquake prediction by animals: Evolution and sensory perception. Bull. Seismol. Soc. Am. 2000;90:312–323.

26. Morton IL. Headaches prior to earthquakes. Int. J. Biometeorol.1988;32:147–148.

27. Rose MS, Verhoef MJ, Ramcharan S. The relationship between chinook conditions and women's illness-related behaviours. Int. J. Biometeorol. 1995;38:156–160.

28. Piorecky J, Becker WJ, Rose MS. Effect of chinook winds on the probability of migraine headache occurrence. Headache.1997;37:153–158.

29. Balk M, Bose M, Ertem G, Rogoff DA, Rothschild LJ, Freund FT. Oxidation of water to hydrogen peroxide at the rock-water interface due to stress-activated electric currents in rocks. Earth Planet. Sci. Lett. 2009;283:87–92.

30. nan ST, Akgül C, Seyis R, Saatçılar S, Baykut S, Ergintav S, Ba M. Geochemical monitoring in the Marmara region (NW Turkey): A search for precursors of seismic activity. J Geophys Res. 2008;113:B03401. doi: 10.1029/2007JB005206.

31. Pérez NM, Hernández-del-Valle G, Igarashi G, Trujillo I, Nakai S, Sumino H, Wakita H. Searching and detecting earthquake geochemical precursors in CO_2-rich groundwaters from Galicia, Spain. Geochem. J. 2008;42:75–83.

32. Biagi PF, Piccolo R, Ermini A, Fujinawa Y, Kingsley SP, Khatkevich YM, Gordeev EI. Hydrogeochemical precursors of strong earthquakes in Kamchatka: further analysis. Nat. Hazards Earth Syst. Sci. 2001;1:9–14.

33. Balderer W, Leuenberger F. Effects of the Cinarcik-Ismit August 17, 1999 earthquake on the composition of thermal and mineral waters as revealed by chemical and isotope investigations. Geophys. Int. 2002;41:385–391.

34. Balderer W, Leuenberger F. Observation of Fluorescence Spectra of Ground Water in Areas of Tectonic Activity: Could It Act as a Precursor? In: Sen P, Das NK, editors. Geochemical Precursors for Earthquakes. McMillan; Kolkata, India: 2006. pp. 22–30.

35. Michetti AM. Active tectonics and seismic hazard in the Central Western Southern Alps: A review. Geophys. Rese. Abstr.2005;7:10830.

36. Saran M, Summer KH. Assaying for hydroxyl radicals: Hydroxylated terephthalate is a superior fluorescence marker than hydroxylated benzoate. Free Radic. Res. 1999;31:429–436.

37. Singh RP, Kumar JS, Zlotnicki J, Kafatos M. Satellite detection of carbon monoxide emission prior to the gujarat earthquake of 26 January 2001. Appl. Geochem. 2010;25:585–580.

38. Milne J, Lee AW. Earthquakes and Other Earth Movements. 7th ed. Kegan Paul, Trench, Trubner & Co; London, UK: 1939. p. 242.

39. Turcotte DL. Earthquake prediction. Annu. Rev. Earth Planet. Sci. 1991;19:263–281.

40. Ikeya M, Furuta H, Kajiwara N, Anzai H. Ground electric field effects on rats and sparrows: Seismic anomalous animal behaviors (SAABs) Jpn. J. Appl. Phys. 1996;35:4587–4594.

41. Wang K, Chen QF, Sun S, Wang A. Predicting the 1975 Haicheng Earthquake. Bull. Seismol. Soc. Am. 2006;96:757–795.

42. Zhu F, Wu G. Haicheng Earthquake. Seismological Press; Beijing, China: 1982. p. 220. (in Chinese)

43. Grant RA, Halliday T. Predicting the unpredictable: Evidence of pre-seismic anticipatory behaviour in the common toad. J. Zool.2010;281:263–271.

44. Gittins SP, Parker AG, Slater FM. Population characteristics of the common toad (Bufo bufo) visiting a breeding site in mid-Wales.J. Anim. Ecol. 1980;49:161–173.

45. Fidani C. The earthquake lights (EQL) of the 6 April 2009 Aquila earthquake in Central Italy. Nat. Hazards Earth Syst. Sci.2010;10:967–978.

46. Serpieri A. Nuove osservazioni sul terremoto avvenuto in Italia il 12 marzo 1873 e riflessioni sul presentimento degli animali per i terremoti. Rendiconti del R. Istituto lombardo. 1873;6:25–33.

47. Reading CJ, Clarke RT. Male breeding behaviour and mate acquisition in the common toad, Bufo bufo. J. Zool. 1983;201:237–246.

48. Liu JY, Chuo YJ, Shan SJ, Tsai YB, Chen YI, Pulinets SA, Yu SB. Pre-earthquake ionospheric anomalies. Ann. Geophys.2004;22:1585–1593.

49. Chen YI, Liu JY, Tsai YB, Chen CS. Statistical test for pre-earthquake ionospheric anomaly. Terr. Atmo. Ocean. Sci.2004;15: 385–396.

50. Hayakawa M, Kasahara Y, Nakamura T, Muto F, Horie T, MAekawa S, Hobara Y, Rozhnoi AA, Solivieva M, Molchanov OA. A statistical study on the correlation between lower ionospheric

perturbations as seen by subionospheric VLF/LF propagation and earthquakes. J. Geophys. Res. 2010;115: A09305.

51. Hayakawa M, Kasahara Y, Hobara TNY, Rozhnoi A, Solovieva M, Molchanov OA. On the correlation between ionospheric perturbations as detected by subionospheric VLF/LF signals and earthquakes as characterized by seismic intensity. J. Atmos. Sol.-Terr. Phys. 2010;72: 982–987.

52. Tsolis GS, Xenos TD. A qualitative study of the seismo-ionospheric precursors prior to the 6 April 2009 earthquake in L'Aquila, Italy. Nat. Hazards Earth Syst. Sci. 2010;10:133–137.

53. Eftaxias K, Balasis G, Contoyiannis Y, Papadimitriou C, Kalimeri M, Athanasopoulou L, Nikolopoulos S, Kopanas J, Antonopoulos G, Nomicos C. Unfolding the procedure of characterizing recorded ultra-low frequency, kHZ and MHz electromagetic anomalies prior to the L'Aquila earthquake as pre-seismic ones—Part II. Nat. Hazards Earth Syst. Sci. 2010;10: 275–294.

54. Eftaxias K, Athanasopoulou L, Balasis G, Kalimeri M, Nikolopoulos S, Contoyiannis Y, Kopanas J, Antonopoulos G, Nomicos C. Unfolding the procedure of characterizing recorded ultra-low frequency, kHZ and MHz electromagetic anomalies prior to the L'Aquila earthquake as pre-seismic ones—Part I. Nat. Hazards Earth Syst. Sci. 2009;9:1953–1971.

55. Fidani C. Electromagnetic Signals Recorded by Perugia and S. Procolo (Fermo) Stations before the Aquila Earthquakes. GNGTS-Gruppo Nazionale di Geofisica della Terra Solida; Trieste, Italy: 2009. pp. 370–373.

56. Derr JS, St-Laurent F, Freund FT, Thériault R. Earthquake Lights. In: Gupta H, editor. Encyclopedia of Solid Earth Geophysics. Springer; Dordrecht, The Netherlands: 2010.

57. Derr JS. Earthquake lights: A review of observations and present theories. Bull. Seismol. Soc. Am. 1973;63:2177–2187.

58. Buskirk RE, Frohlich CL, Latham GV. Unusual animal behavior before earthquakes: A review of possible sensory mechanisms. Rev. Geophys. 1981;19: 247–270.

59. Ikeya M. Earthquakes and Animals: From Folk Legends to Science. World Scientific; London, UK: 2004. p. 295.

60. Ikeya M, Yamanaka C, Mattsuda T, Sasaoka H, Ochiai H, Huang Q, Ohtani N, Komuranani T, Ohta M, Ohno Y, Nakagawa T. Electromagnetic pulses generated by compression of granitic rocks and animal behavior. Episodes. 2000;23:262–265.

61. Witze A. The sleeping dragon. Nature. 2009;459:153–157.

62. World Stress Map Project Available online: http://dc-app3-14. gfz-potsdam.de/ (accessed on 1 June 2011)

63. King BV, Freund F. Surface charges and subsurface space charge distribution in magnesium oxide containing dissolved traces of water. Phys. Rev. 1984;29B:5814–5824.

64. Hayakawa M, Hattori K, Ohta K. Monitoring of ULF (ultra-low-frequency) geomagnetic variations associated with earthquakes. Sensors. 2007;7:1108–1122.

65. Molchanov OA, Schekotovm A, Federov E, Belyaev G, Gordeev EI. Preseismic ULF electromagnetic effect from observation at Kamchatka. Nat. Hazards Earth Syst. Sci.2003;3:203–209.

66. Hayakawa M, Ito T, Hattori K, Yumoto K. ULF electromagnetic precursors for an earthquake at Biak, Indonesia on February 17, 1996. Geophys. Res. Lett. 2000;27:1531–1534.

67. Rycroft MJ, Harrison RG, Nicoll KA, Mareev EA. An overview of earth's global electric circuit and atmospheric conductivity.Space Sci. Rev. 2008;137:83–105.

68. Zhao B, Wang M, Yu T, Guirong X, Wan W, Liu L. Ionospheric total electron content variations prior to the 2008 Wenchuan Earthquake. Int. J. Remote Sens. 2010;31:3545–3557.

69. Liu JY, Chen CH, Chen YI, Yang WH, Oyama KI, Kuo KW. A statistical study of ionospheric earthquake precursors monitored by using equatorial ionization anomaly of GPS TEC in Taiwan during 2001–2007. J. Asian Earth Sci. 2010;39:76–80.

70. Liu JY, Chuo Y, Shan S, Tsai Y, Chen Y, Pulinets S, Yu SB. Pre-earthquake ionospheric anomalies registered by continuous GPS TEC measurements. Ann. Geophys. 2004;22:1585–1593.

71. Livingstone DR. Oxidative stress in aquatic organisms in relation to pollution and aquaculture. Revue Méd. Vét.2003;154:427–430.

72. Lushchack VI. Environmentally induced oxidative stress in aquatic animals. Aquat. Toxicol. 2011;101:13–30.

73. Fridovich I. Superoxide radical and superoxide dismutases.Annu. Rev. Biochem. 1995;64:97–112.

74. Davies KJ, Delsignore ME. Protein damage and degradation by oxygen radicals. III. Modification of secondary and tertiary structure. J. Biol. Chem. 1987;262:9908–9913.

75. Imlay JA, Linn S. DNA damage and oxygen radical toxicity. Science. 1988;240:1302–1309.

76. Michaelidis B, Ouzounis C, Paleras A, Pörtner HO. Effects of long-term moderate hypercapnia on acid-base balance and growth rate in marine mussels Mytilus galloprovincialis. Mar. Ecol. Prog. Ser. 2005;293:109–118.

77. Reid SD, Dockray JJ, Linton TK, McDonald DG, Wood CM. Effects of chronic environmental acidification and a summer global warming scenario: Protein synthesis in juvenile rainbow trout (Oncorhynchus mykiss) Can. J. Fish. Aquat. Sci. 1997;54:2014–2024.

78. Amitai Y, Zlotogorski Z, Golan-Katzav V, Wexler A, Gross D. Neuropsychological impairment from acute low-level exposure to carbon monoxide. Arch. Neurol. 1998;55:845–848.

79. Feder ME, Burggren WW. Environmental Physiology of the Amphibians. University of Chicago Press; Chicago, IL, USA: 1992.

80. Vitt LJ, Caldwell JP, Wilbur HM, Smith DC. Amphibians as harbingers of decay. BioScience. 1990;40:418.

81. Valko M, Morris H, Cronin MT. Metal, toxicity and oxidative stress. Curr. Med. Chem. 2005;12:1161–1208.

82. Mahapatra PK, Mohanty-Hejmadi P, Chainy GB. Specific limb abnormalities induced by hydrogen peroxide in tadpoles of Indian jumping frog, *Polypedates maculatus*. Indian J. Exp. Biol.2001;39:1103–1106.

83. Sadinski WJ, Dunson WA. A multilevel study of effects of low pH on amphibians of temporary ponds. J. Herpetol. 1992;26:413–422.

84. Alvarado RH, Cox TC. Action of polyvalent cations on sodium transport across skin of larval and adult Rana catesbeiana. J. Exp. Zool. 1985;236:127–136.

85. Dobrovolsky IP, Zubkov SI, Miachkin VI. Estimation of the size of earthquake preparation zones. Pure Appl. Geophys.1979;117:1025–1044.

FE Based Vulnerability Assessment of Highway Bridges Exposed to Moderate Seismic Hazard

C. Mullen[1]

[1] Department of Civil Engineering, University of Mississippi-Oxford, USA

INTRODUCTION

The assessment of seismic vulnerability in regions where the risk from earthquake shaking is considered moderate poses special problems in terms of establishing critical conditions for failure and the importance and urgency for taking action. Research studies sponsored at the University of Mississippi (UM) over a period of about 10 years by the Mississippi Emergency Management Agency (MEMA) and Mississippi

Department of Transportation (MDOT), respectively, have been aimed at identifying the vulnerability of select critical highway bridges subject to significant ground shaking from the New Madrid Seismic Zone (NMSZ).

The historical occurrence of multiple but infrequent major seismic events in the NMSZ exceeding seismic moment of M 7 has been established by geophysicists and seismologists through numerous surveys of surface rupture features and pale seismological excavations conducted throughout the region (for example, see [10, 18]). Planners in both state and federal agencies are concerned about the consequences of both physical and economic damage posed by the next major recurrence of a potentially catastrophic earthquake along the fault. The United States (US) Federal Emergency Management Agency (FEMA) sponsored a major research study [3] to investigate the multi-state regional consequences of a hypothetical event of M 7.7 on both buildings and bridges. The bridges in Mississippi discussed in this chapter represent critical lifelines exposed to the earthquake threat that are located along the evacuation routes and economic supply chains for communities in the northern part of the state as well as the tri-state metropolitan area of the city of Memphis, Tennessee, having population of about 1.3 million.

A myriad of uncertainties exist for both the rare but potentially catastrophic seismic events and the multiple factors affecting the response of these soil-foundation-structure systems. In the absence of ground motion records for the severe historical events in the seismic zone under consideration, a simulation based approach is adopted to highlight the salient features of both the input and response at the site. The vulnerability assessment requires that reasonable behavioral response and multiple failure limit states be examined under a range of possible ground motion intensities. While a probabilistic approach is desirable overall, a deterministic approach enables the examination of the key response characteristics and the detailed information needed to establish relative importance of different limit states including soil capacity, pile/column axial and flexural strength, and member/system instability.

The bridge seismic vulnerability studies in this chapter highlight the challenges posed by the need to balance the level of sophistication of the finite element (FE) simulation with the:

- state of knowledge of the bridge facilities, their seismic exposure, and local site conditions
- project objectives in order to provide safe and economic decision making for hazard mitigation and emergency response and mobilization planning

Lessons learned and discussed herein are the result of over a decade of research at UM sponsored at the multidisciplinary Center for Community Earthquake Preparedness (CCEP) and graduate level studies by a number of students supported by the Department of Civil Engineering.

SEISMIC HAZARD AND INVENTORY CHARACTERIZATION FOR THE STUDY REGION

In [10] the 1811, 1812 sequence of three distinct earthquakes corresponding to rupture along separate segments of the irregular shaped New Madrid fault is described. More recently, in the FEMA study [3], a scenario established for emergency planning purposes comprising a single M7.7 event consisting of sequential rupture along all three segments. Seismicity of smaller events recorded using a strong motion instrument array during an almost 30 year span is plotted in Figure 1 to which the approximate location of the study region has been added.

According to the 2012 data compiled for the National Bridge Inventory (NBI) [6] in the US by the Federal Highway Administration (FHWA), a total of 18,459 highway bridges are found in the 82 counties in the state of Mississippi (MS). The study region contains only a small subset of this inventory and may be approximately characterized as the counties located in north MS most likely to experience moderate ground shaking from a major event in the NMSZ. Based on default inventory data contained in the GIS-based software, Hazards US-Multihazard (HAZUS-MH) [5] created under sponsorship by FEMA for use in emergency management planning, 1133 bridges are exposed to the moderate seismic hazard.

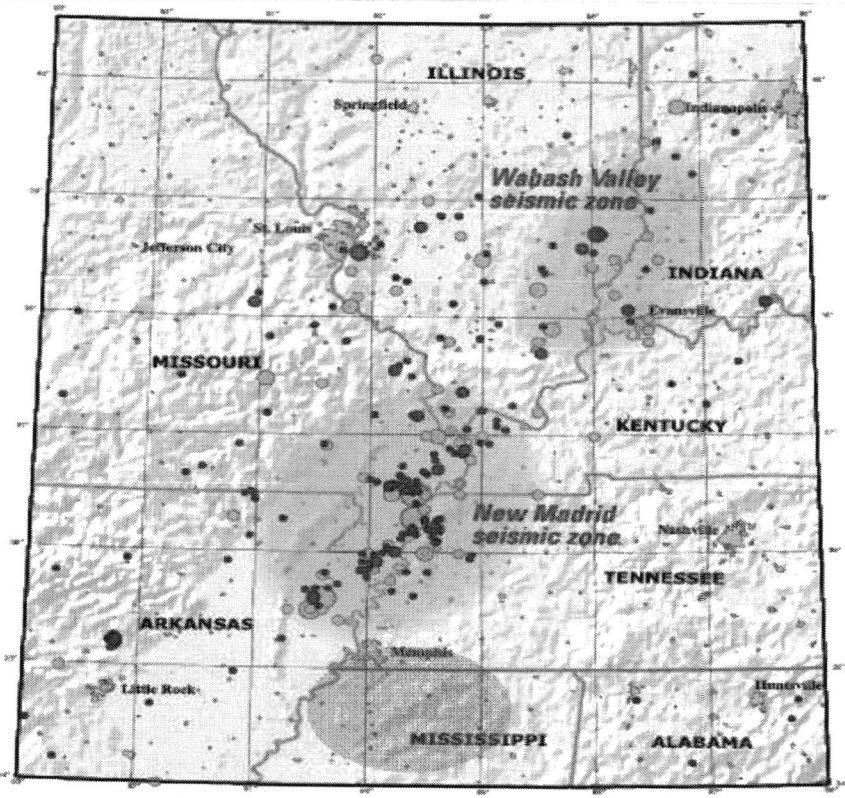

Figure 1: Recent seismicity in NMSZ and surrounding multi-state region exposed to risk of a repeat of historic catastrophic events (M7-M8); red circles give epicenters for events > M2.5 during the period, 1974-2002 [from 19]; the study region is represented by the blue shaded area.

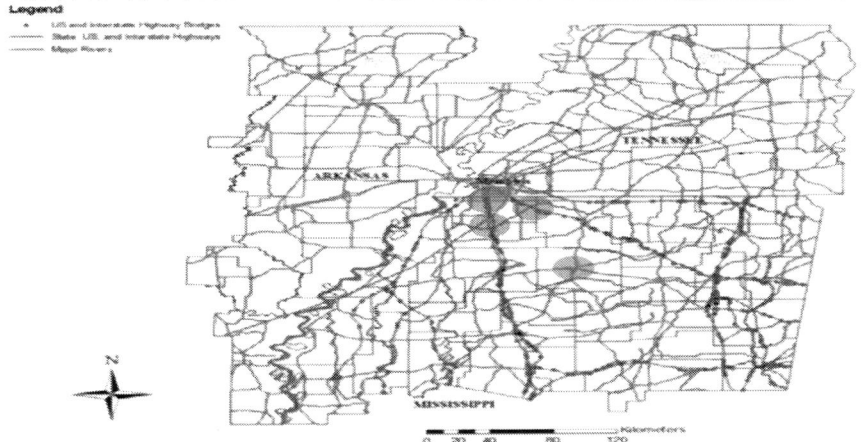

Figure 2: NBI bridge inventory in study region shown by open circles; green circles show bridges located in north Mississippi on the major access routes for the Memphis metropolitan area, those investigated using FE based seismic vulnerability analysis lie within shaded areas; red lines indicate highways on federal and state system; blue lines represent major rivers to show critical water crossings

The seismic vulnerability of all bridges in MS has been examined from a risk or loss estimation point of view in both [3] and [13]. In each study, the HAZUS-MH methodology has been implemented which depends on use of fragility curves assigned to bridge classes included in the NBI system. No study has yet been performed to assess seismic vulnerability using FE as the basis of the loss estimation. The present study provides a first step toward such a more comprehensive study and focuses on five bridges at a variety of sites in the study region investigated during three separate projects. Figure 2 shows the locations of the sites in relation to the NBI inventory supplied in HAZUS-MH and federal and state highway system.

The select bridges studied have been modeled to varying degrees of complexity with both two-dimensional (2D) and three-dimensional (3D) computational simulations including eigenvalue, linear dynamic, nonlinear static, and nonlinear dynamic. Earthquake time histories have been generated to capture a range of intensities from M6 to M8 and peak ground accelerations (PGA) in the approximate range, PGA=5-50% g, depending on the site, study objectives, and methodology. It is noteworthy that, over the period of the studies, significant changes have occurred in the understanding of the earthquake risk and level of ground shaking to be expected. Each study used the best available knowledge at the time. The earliest study was performed for MEMA in the context of a broader study of the seismic vulnerability of facilities located on the UM campus [17]. The motivation for the study was the belief that the University was and remains a key to economic development in the state as well as a place of both historic value and a population center of relatively high density. The study included an approximately 70 year old bridge that serves as a major entrance as seen in Figure 3. The bridge was designed by MDOT prior to any recognition of a significant seismic threat in the applicable design code. This bridge served as the first attempt at a detailed 3D FE-based evaluation of seismic response and vulnerability assessment [12]. The evaluation was performed with both fixed base and soil-foundation-substructure interaction boundary conditions to capture the influence of high embankments on the response of the structural components with emphasis on the pier columns.

Figure 3: East Gate Bridge carrying traffic from University Avenue in the City of Oxford at the entrance to the UM main campus [17]; present day bridge, old private railroad tracks, and right-of-way have been replaced by a modern city roadway.

A second study [14] was performed for MDOT for what the Bridge Division deemed a critical facility that provides access from a major interstate highway to a vital economic development region in the state. The region is located within the fastest growing county in the state and one of the fastest growing in the nation due to its proximity to the metropolitan area of the city of Memphis, Tennessee. Approximately 30 years old, this bridge shown in Figure 4 was built to low seismic standards. The code recognized by MDOT was and remains the one published by the American Association of State Highway and Transportation Officials (AASHTO). Even when the first edition of the AASHTO Bridge Load and Resistance Factor Design (LRFD) Specifications appeared in 1994, the ground motion demand at the site was only about PGA=0.15g.

The third and most recent study was performed for MEMA to investigate findings of the FEMA sponsored NMSZ catastrophic earthquake study [3] on the impact of an M 7.7 event on bridges in MS. Using a HAZUS-MH fragility curve based analysis which estimated conditional probabilities of damage at four basic limit states (slight, moderate, extensive, and complete), the study found that only six bridges in the entire state would have a significant probability exceeding slight damage. The purpose of the FEMA study was to provide states affected by the NMSZ a basis for establishing earthquake components of their federally mandated mitigation plans. The MEMA study used an FE based approach to establish vulnerability considering more site and facility specific information. In consultation with MDOT personnel, three bridges shown in Figure 5 were identified for study. All are located on major evacuation/mobilization routes which crossed the Coldwater River. The bridges were deemed near the edge of significant ground shaking based on the FEMA study. The rationale was that if these showed evidence of significant vulnerability then bridges closer to the NMSZ would then be at similar or higher risk.

Figure 4: Bridge carrying *MS* 302 over Interstate highway *I* 55; (left) looking toward Southaven, MS, a fast growing city forming part of the metropolitan area of the City of Memphis; (right) view of the intermediate bents and girders of the two closely spaced bridges.

Figure 5: Three bridges crossing the Coldwater River on lifelines serving the study area; (left) view of southbound *I* 55 bridge; (middle) nearby *US* 51 bridge showing piled bents carrying simple spans; (right) view of northbound *US* 78 bridge.

GROUND MOTION SIMULATION FOR THE STUDY SITES

The lack of seismic records of significant earthquake events in the NMSZ makes the task of selecting ground motion excitation for response analysis a challenge. The state of knowledge of the causative features of the fault and the expected attenuation of motions from the source has changed over time and remains an area of significant debate and research. Spectral physics-based parametric source and attenuation models have provided a rational basis for the case studies presented here.

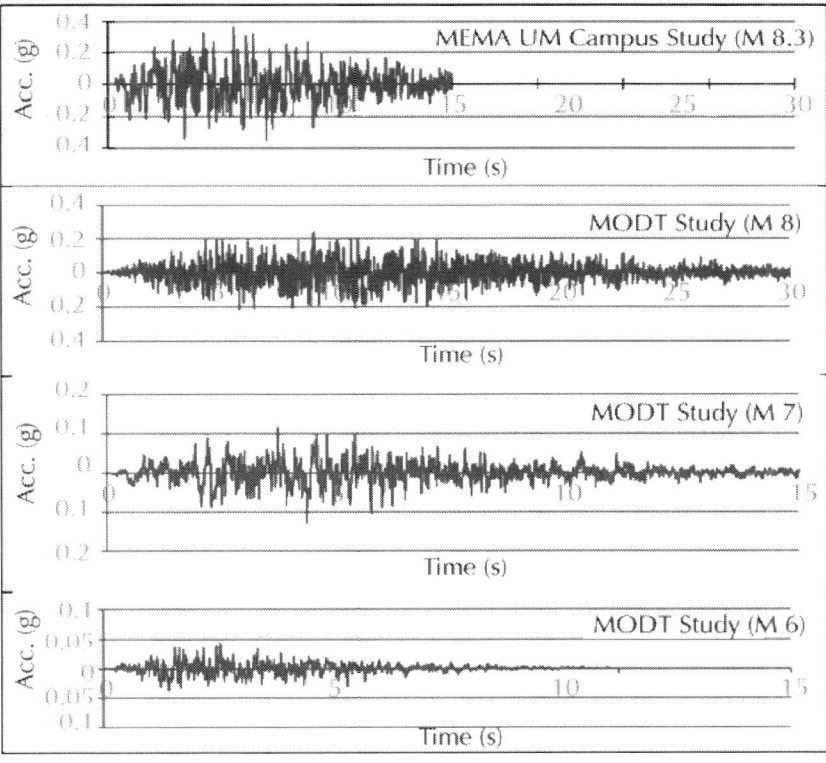

Figure 6: resultant horizontal ground acceleration time histories used in FE model analyses; MEMA UM campus study [17]; MDOT study [14]; MEMA Coldwater River bridges study [16]

Figure 6 shows resultant horizontal ground motion realizations generated for the various studies assuming 2D propagation from an assumed epicenter usually taken as Marked Tree, Arkansas, the town nearest to the southernmost position of the New Madrid fault. In the MEMA campus and MDOT bridge studies, orientation of the bridge was considered and component motions were then extracted for application to the FE models. In the absence of a 3D propagation model, requiring definition of layered media in a spherical coordinate system, vertical motion was obtained by uniformly scaling the resultant horizontal motion by the commonly assumed factor of 2/3.

In the MEMA UM campus study [17], the input horizontal motion realization for the UM campus was generated by others for M6.3 and M 8.3 events having source along the nearest (southernmost) segment of the NMSZ. In the MDOT bridge study [14], software was obtained from the US Geological Survey (USGS) and source model parameters and attenuation relations were identified in consultation with USGS and the University of Memphis enabling simulation of multiple realizations at arbitrary intensities. Events of nominal M 6, 7, and 8 were selected in order to capture different response levels. In the MEMA Coldwater River bridges study [16], a FEMA scenario of M 7.7 was adopted to be consistent with their results for the multi-state NMSZ region which were based on a distributed source model involving slip along the entire southern segment of the New Madrid fault. Since the study provided only PGA contours, not time histories, the MDOT study realization for the M8 scenario case (Fig. 6) was scaled to achieve an input motion with PGA corresponding to that of the M 7.7 scenario at the bridge site locations (approximately PGA=0.25g).

Source spectral models for the very large intensity events were such that all ground motions have significant energy in the 1-2 Hz range coinciding with fundamental and low natural frequencies of the bridges.

FE MODELING OPTIONS

When using FE as the basis of vulnerability assessments, it is important to make several basic decisions regarding modeling approach including probabilistic versus deterministic and simple versus complex. These choices influence at the most general level, the software to be used, and at the most specific level, the key modeling assumptions such as system scope, boundary conditions, incorporation of soil-structure interaction (SSI), and focus on lumped parameter, 2D structural, or 3D continuum finite elements. Rather than propose a comprehensive view on the proper choices for all possible objectives, the select bridge study cases are offered as the possible range one might consider.

In the MEMA UM campus study [17], no prior knowledge existed. As a result of this uncertainty about what might be expected as well as a strong desire to ensure the safety of the many thousands of students, employees, and visitors to the campus and a major concern about the impact of significant losses to the future functioning of the university enterprise and consequential economic impacts on the state, the sponsors sought the most realistic view possible given the state of the art at the time. In response to this objective, the analysts committed to full 3D nonlinear dynamic FE simulation including SSI in cases where it might have a significant influence on the response. The project was initiated in the mid-1990s when the software ABAQUS [7] provided many desirable features including 3D nonlinear beam-column (structural) elements (B33) with user input moment-curvature relations and 3D continuum (solid) "infinite" elements (CIN3D8) with shape functions capturing radiation damping, in effect providing non-reflecting boundaries which allow dissipation of wave energy propagating radially away from the FE model.

There was little experience with the modeling approach at the time of the study and no experience with the nonlinear beam-column and radiation damping elements, so validation analyses were performed [9]. Detailed drawings were available from the bridge designer (MDOT), and a series of detailed models were developed to establish confidence in each subsequent level of complexity. Static self-weight analysis was first performed using a so-called fixed-based model (no soil stiffness included) to represent structural connectivity and weight and stiffness characteristics. Basic features of the fixed base model are shown in Fig. 7.

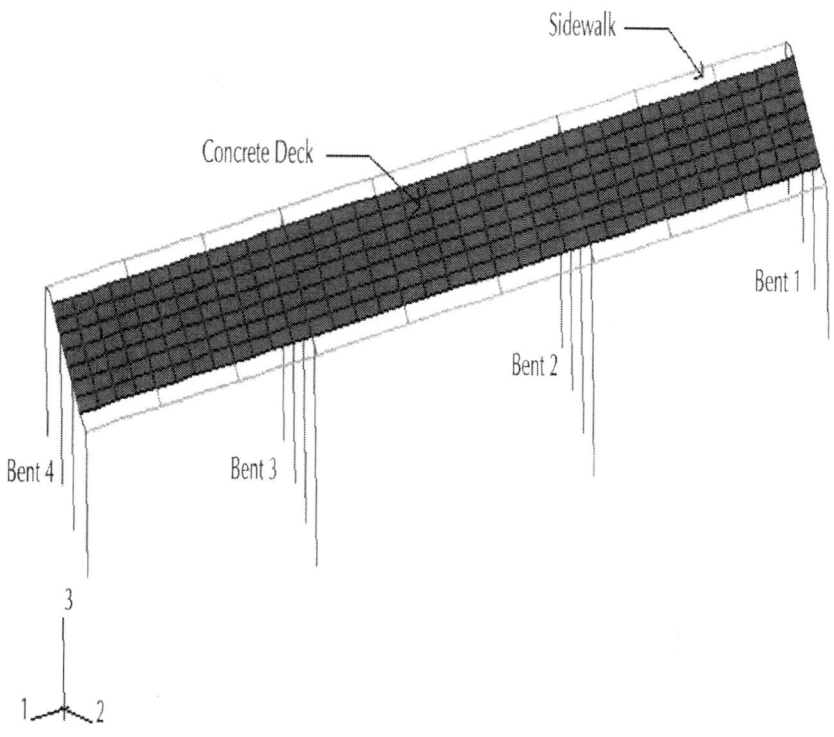

Figure 7: Fixed-base FE model of East Gate Bridge for MEMA UM campus study[17]; bents modeled with nonlinear beam-column elements; composite concrete deck-steel girder superstructure modeled using concrete plate elements for deck and linear beam elements for steel girders; no soil degrees-of-freedom.

Once an acceptable result was obtained from the static analysis, an eigenvalue analysis was performed to estimate structural mass distribution characteristics and associated mode shapes and frequencies. Since the ground motions shown in Figure 6 accounted primarily for propagation through the earth's crust, modification and possible amplification as the seismic waves propagated through soil at the bridge site was not considered. To account for this limitation, a one-dimensional (1D) vertical wave propagation analysis [12, 17] was performed using a model of the top 100 ft of soil layers based on data obtained from soil borings. The analysis incorporated nonlinear softening of dynamic shear moduli at high strains and enabled generation of input motions to all fixed degrees-of-freedom (DOF) in the FE model regardless of elevation, in this case, at both the base of the columns of the intermediate bents and the level of the end abutment pile caps.

As Figure 3 shows, there is a significant difference (over 30 ft) in elevation between the abutments and the intermediate bents. Furthermore, the deck girders are built into concrete end walls where fill material is placed beneath the roadway. Between the abutments and what is now a roadway, steep embankments are found. To incorporate the interaction between the soil immediately below the footings of the intermediate piers, the embankments, and the structural system, the significantly more elaborate model shown in Figure 8 was developed [12,17].

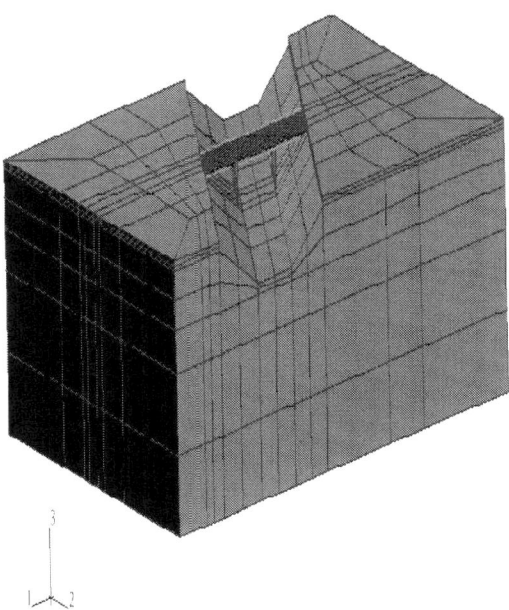

Figure 8: Subsurface geology and embankment interaction FE model of East Gate Bridge for MEMA UM campus study [12, 17]; end walls modeled with shell elements; active/passive soil pressure resistance modeled with nonlinear springs connecting end wall and back fill soil elements; embankment soil and subsurface geology modeled with elastic 3D solid elements; radiation damping at absorbing boundaries modeled with 3D solid infinite elements.

The MDOT study was the first earthquake vulnerability study performed in the state for its Bridge Division. Again because of the many uncertainties, a 3D detailed FE based simulation approach [15] was adopted to provide the most accurate estimate of likely response. The bridge system was much larger than the one in the UM campus study due to the overcrossing of an interstate highway which now carries three lanes of traffic in each direction and the presence of two bridge frame substructures separated by a only a small gap between bents (see Fig. 4). The servicing of a large commercial center and a rapidly growing residential community required the bridge to carry a total of nine lanes of traffic, each substructure carrying traffic in one of the two directions. Embankments again created a significant difference in elevation of approximately 20 ft between soil beneath respective roadway pavements, but here the embankments were sloped to accommodate access to/from the interstate highway.

As shown in Figure 9, there were four continuous deck spans totaling approximately 350 ft. The substructures now included both piled footings at the end abutments and central intermediate bent and spread footings at the two other intermediate bents. A low-rise building SSI study [9] had demonstrated the importance of including a refined mesh locally around spread footings to account for soil softening under large seismic shaking. The detail view in Figure 9 shows the refinement pattern used around the bridge footings.

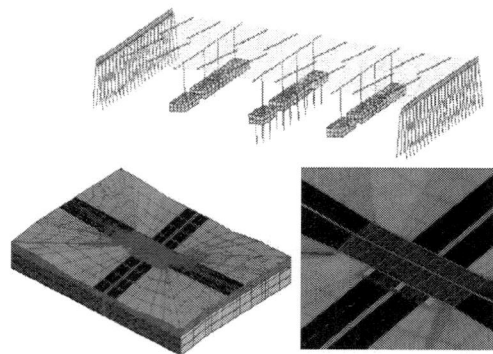

Figure 9: Subsurface geology and embankment SSI FE model of I55/MS302 Goodman Road Overcrossing for MDOT study [14]; concrete girders and bent frame members modeled with 3D nonlinear beam elements; concrete

deck and footing modeled with shell elements (top figure shows soil elements connecting to footing shell elements); soil modeled using 3D solid elements with a Drucker-Prager cap material model for nonlinear response at high strains; radiation damping at absorbing boundaries modeled with 3D solid infinite elements

The MEMA Coldwater River bridges study [16] was originally intended to support a multi-state regional (National Level) earthquake Exercise (NLE) sponsored by FEMA with participation by MEMA. A major flood along the MS River threatened to overtop the levees protecting the farming communities in the Delta region, so MEMA personnel were called away from the exercise, and the input from the bridge study was not required as planned. The long term objective of the study to assess the bridge vulnerability was nonetheless pursued but without as much urgency.

The three Coldwater River bridges consisted of multiple intermediate bents (up to 42 in one case) supporting composite concrete deck slabs over short simple spans (40-50 ft) and a longer central span (100-120 ft) over the main navigable channel. The deck in the central span was usually continuous over several adjacent spans and consisted of a multi-cell concrete box girder or a composite concrete steel girder section. With a limited budget and time frame, a 3D model of the entire bridge with SSI was not attempted. A simpler approach was taken that focused on characterizing the main perceived sources of vulnerability.

Again, design drawings were available from MDOT along with soil borings and test pile logs. The drawings indicated the structures had been built in the 1950s and 1960s, and lacked any consideration of seismic loading in the design. The location of the bents in the flood plain of the river with, in several cases, soil in the top layer permanently saturated, allowed the possibility of weak lateral resistance of the soil and liquefaction under strong ground shaking. The modeling approach thus focused on 2D representation of lateral resistance of the typical intermediate bents in each bridge and 3D representation of the continuous span box girders.

Figure 5 shows that the intermediate bents consist of 4-5 relatively short concrete piles with batters on the outer piles tied together by a concrete pile cap that support bearings for the deck girders. Figure 10shows the representation of this structural system as modeled in the SAP2000 software [1]. The piles in this system were designed for vertical (deck weight and vehicle live) loads primarily, so the potential vulnerability is from lateral inertial load generated by seismic shaking. Under lateral forces, the piles have a tendency to bend under the lateral resistance from the soil. Furthermore, the overturning moment associated with the deck lateral load develops increased compressive axial loads in the outer (batter) piles far in excess of their design assumptions.

Key aspects of the modeling are the axial and bending capacity of the concrete pile section, the lateral stiffness of the soil, the unsupported length of the pile, and the depth of pile embedment. In keeping with the simplified assumptions, linear vertical and horizontal soil springs were used to represent the soil resistance. Surprisingly, standard geotechnical and bridge engineering textbooks and even some advanced earthquake engineering ones offer little on methods to determine the stiffness properties of soil, choosing to focus rather exclusively on capacity estimation. Results presented in a FEMA guidance document [4] were used to estimate the spring constants considering the projected area of the pile and the elastic modulus of the soil.

Isolation of the intermediate bents for lateral load analysis is valid to the extent that the deck moves uniformly so that no bending or torsional resistance is provided by adjacent bents. The simple deck spans help to minimize this effect through the discontinuity of the bearings. In the case of continuous main spans, however, the deck is supported on pile supported concrete piers with either one or two columns of significantly different heights and size, so significant resistance from adjacent bents is anticipated. Figure 11 shows a 3D model developed using another FE software [2] oriented toward bridge design analysis used to explore the effect of the interaction between bents in these spans.

FE EVALUATION PROCESS – SYSTEM BEHAVIOR ANALYSIS

The previous section indicates that the goals of the vulnerability evaluation influences the selection of FE modeling options including software (structural or general purpose), level of analysis (2D or 3D), element selection (structural or continuum), connectivity (rigid connections or flexible bearings), boundary conditions (fixed, flexible, or absorbing). These choices not only influence the behavior and response details that may be estimated and visualized, they also determine what output measures are available for estimating physical damage, performance characteristics, and vulnerability.

In the MEMA UM campus study [17], a basic analysis approach was established that was followed throughout all the studies. Before proceeding to the complex nonlinear dynamic time history analysis, linear static and eigenvalue preliminary analyses were first performed. The linear static gravity load analysis requires processing of all parameters and procedures involved in estimating the stiffness properties of the system. It is relatively fast computationally and enables visual and quantitative confirmation of element connectivity and effect of support fixity (fixed base models), support flexibility (soil springs), or absorbing boundary conditions (SSI models). The eigenvalue analysis requires processing of all parameters and procedures involved in estimating the mass properties of the system. The analysis is also relatively fast computationally and yields mode shapes and frequencies. These modal properties provide insight into the expected dynamic response characteristics under earthquake loading.

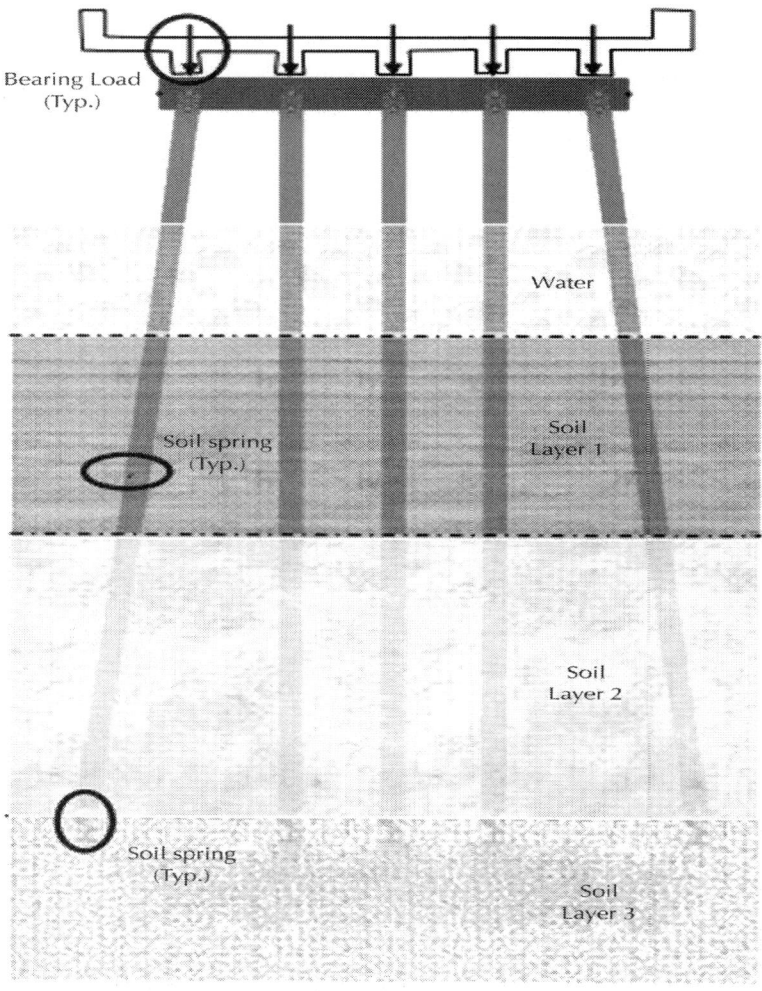

Figure 10: Model of typical intermediate bent for Coldwater River bridge [16] carrying two lanes of interstate highway traffic; concrete piles and cap modeled as frame elements (section behavior modeled with axial-bending interaction using fiber model); ground motion applied to soil springs.

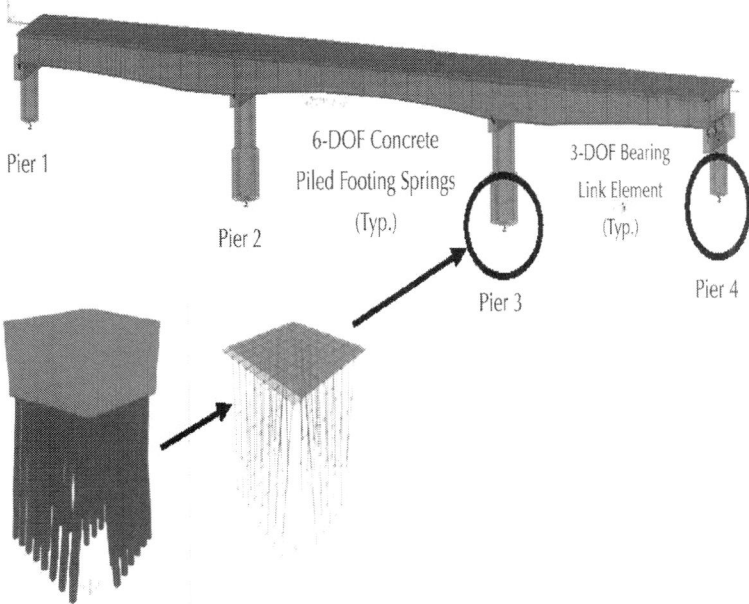

Figure 11: Model [16] of typical 3-span continuous concrete box girder bent for Coldwater River bridge carrying two lanes of interstate highway traffic; concrete pier columns and footing piles modeled with frame elements; box girder flanges and webs and footing pile cap modeled with shell elements; footings modeled with equivalent 6-DOF springs; ground motion applied to footing springs.

Figure 12 illustrates some of the benefits of performing the preliminary analyses before proceeding to the nonlinear time history response analysis. The issues of stiffness and mass distribution become evident from the plotting and animation of the mode shapes associated with global movements of the system. These shapes may be broadly categorized as ones that involve significant net movement of the center of mass of the system and those that do not (sometimes called breathing modes). In the case of the campus bridge shown, it is seen that the mode involving transverse movement of the mass center becomes coupled with a rotational movement because of the skew of the deck necessitated by the angle between the centerlines of the street carried and the one crossed. Also visualized in the case shown in Figure 12 is the effect of the SSI, in this case the embankments and abutments interacting with the main span deck and intermediate bents.

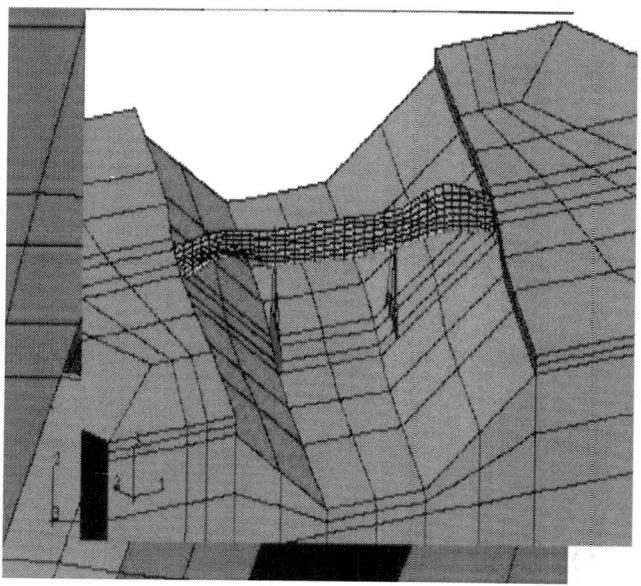

Figure 12: Eigenvalue analysis results for the MEMA UM campus study bridge models [17]; top figures show plan and isometric views of fixed based model transverse mode causing bent and abutment column deformation; bottom figures show comparable modes for SSI model; skew of roadway alignment introduces coupling of translation and rotation of the deck mass as well as bending and torsion of the deck; the resistance provided by the embankment is apparent from the contact developed during rotation.

Behavior similar to that observed for the MEMA UM campus study bridge is found in the case of the MDOT study bridge. Figure 13 shows the transverse mode shape for the fixed base model. The bridge proportions (both deck length to width and deck span to column height ratios) and skew angle are different in the two cases. The translational and rotational coupling is less pronounced, and the transverse column bending is more pronounced.

The eigenvalue analyses not only provide insight regarding the expected deformation patterns, they also provide the frequencies associated with these characteristic modes. These frequencies provide quantitative information which provide insight into the expected influence of the SSI effects as well as the dominance of deformation modes associated with specific earthquake events.

The influence of SSI was examined in detail in the MDOT study which included ambient vibration measurements using a portable array of accelerometers [11, 14]. Simultaneous readings were taken at each bent location under excitation of the bridge by truck traffic. Using a point on the bridge deck as a reference point, frequency response functions were derived that eliminated the influence of the excitation, and system response frequencies were extracted corresponding with excellent correlation to the 3D model SSI case without any model parameter modification. Accelerometers were then moved to the abutments and frequency extraction performed [11] revealing evidence of the participation of the abutments in the transverse mode shape comparable to the one in Figure 13.

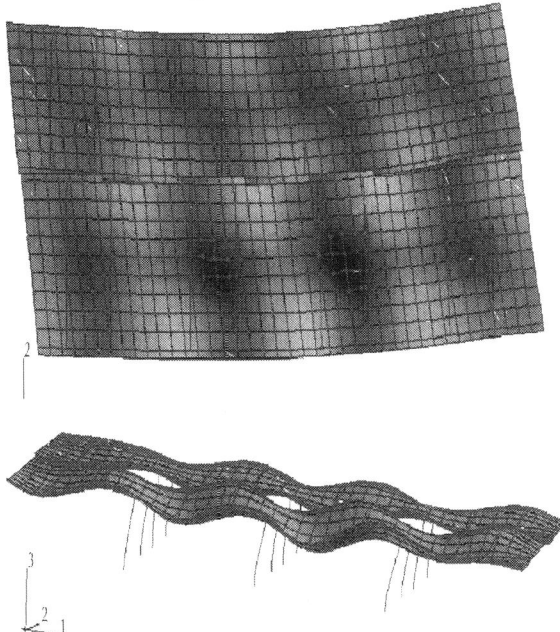

Figure 13: Eigenvalue analysis results for the MDOT bridge study; plan and isometric views of fixed base model transverse mode; lateral deck and column bending dominates response in this pair of adjacent bridge structures; some bending and torsional coupling in the deck is evident.

In the MEMA bridge study, the preliminary analyses were again performed prior to time history analysis. Figure 14 shows that the fundamental mode of vibration for a typical intermediate bent in the interstate highway river crossing is one involving net translation of the deck and corresponding bending of the piles which were designed as axially loaded members. Consideration of the eccentricity of the deck mass with respect to the center of resistance of the soil-pile system provides for expectation of an overturning moment. Such a moment would generate an increase of axial force in one of the batter piles which would combine with the bending action.

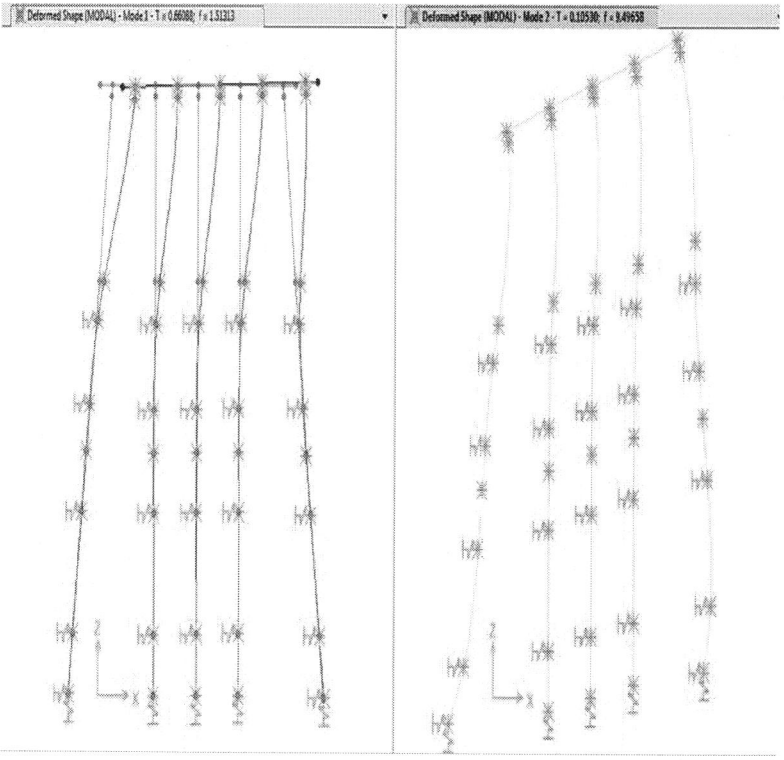

Figure 14: Eigenvalue analysis results for the MEMA bridge study [16]; elevation views of 2D intermediate bent model showing deck transverse displacement (left) and rotation (right) modes; pile bending dominates response although combined action of bending and axial force in the piles is implied.

FE EVALUATION PROCESS — SEISMIC RESPONSE ANALYSIS

The benefit of FE based evaluation is that a great bit of detail of the response of the system is made available through the analysis especially when the time history approach is taken. In essence all DOF selected in the modeling process are accessible over the full length of the simulated event. With further post-processing whether computational or graphical, additional response quantities and behavior can be accessed, plotted, and visualized.

In the MEMA UM campus study [17], it became particularly useful to examine hysteresis of the column section in the plastic hinge region. Figure 15 shows a typical plot of simulated moment-curvature response during the severe (M 8.3) event case. The results demonstrate that the yield limit state is achieved in both directions for a corner column, and the ultimate limit state is achieved in one direction. The latter result provided clear evidence of vulnerability and the possibility of complete failure or collapse.

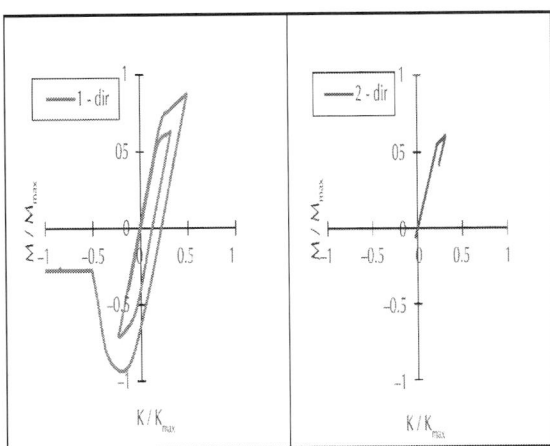

Figure 15: Normalized moment-curvature hysteresis plots for a corner column of the MEMA UM campus study bridge (see Figs. 3, 8, 12) subject to an extreme event (see Fig. 6); nonlinear user-input curves developed separately using a fiber section model [12, 17].

In the MDOT study [14], with the availability of the USGS simulation tool for developing random realizations of input ground acceleration time histories (Fig. 6) at different intensities, limit state determination was enabled for the columns and piles over the full range of damaging events. Comparison of the peak and characteristic responses enabled a performance evaluation of the system based on critical material, section, or member limit states such as first cracking, first yield, plastic hinge formation, and plastic collapse mechanism. In the case of the bridge studied, it was learned that the piles at the abutments and the columns of the central bent provided substantial energy absorption in the extreme event case through ductile hysteretic response in these members. It was also learned that the pile system at the abutments and central bent adequately distributed lateral forces so that the soil remained linear throughout the event. While nonlinear slip at the superstructure to abutment pile cap bearing connection was attempted, this proved too difficult for the software to resolve and convergence was never reached. Ultimately, rigid connections were assumed and the slip mechanism was interpreted as another potential energy absorption source.

In the MEMA Coldwater River bridges study [16], linear dynamic response was performed for most of the analysis runs. An example of the response motion time histories at the level of a typical intermediate bent pile cap is shown in Figure 16. Peak internal force (axial force, shear, and bending moment) responses in the piles were obtained and compared with design values and pile test data.

In the critical case, a nonlinear static pushover analysis was performed to estimate the capacity of the pile system. In the FE model, both geometric and material nonlinear options for the software were used. In the latter option, a fiber representation of the cross-section was used that accounted for 1D nonlinear normal stress-normal strain behavior in the concrete and the steel reinforcement, enabling computation of the force-displacement behavior shown in Figure 17. A nonlinear time history was also run for this critical case.

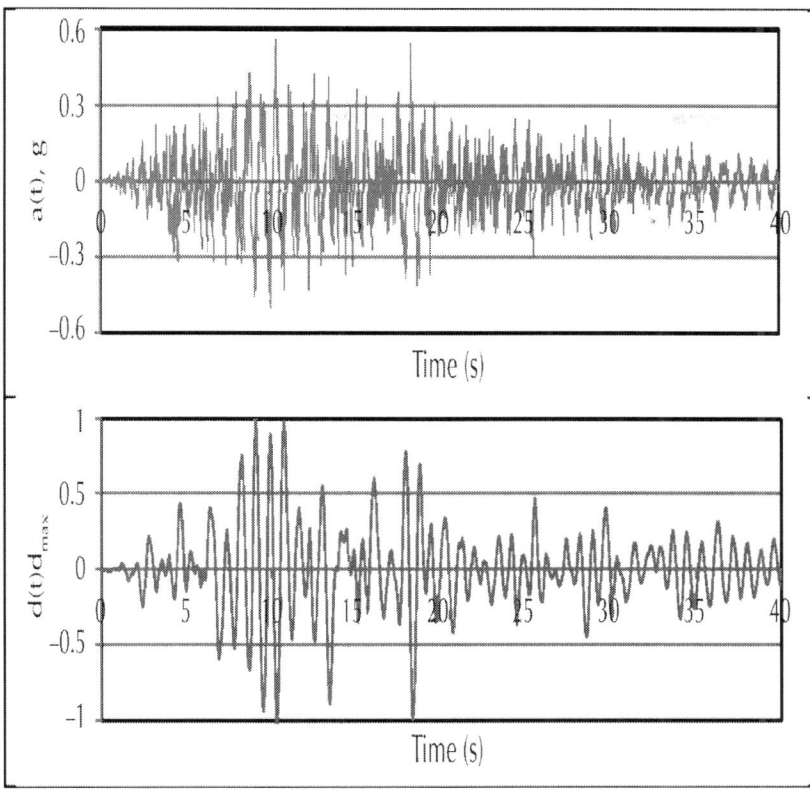

Figure 16: Normalized time history plots of response at center of intermediate bent pile cap for one of the Coldwater River bridges studied [16] (see Figs. 5 and 10) subject to an extreme event (see Fig. 6scaled to PGA=0.25g); top plot shows acceleration; bottom plot shows displacement).

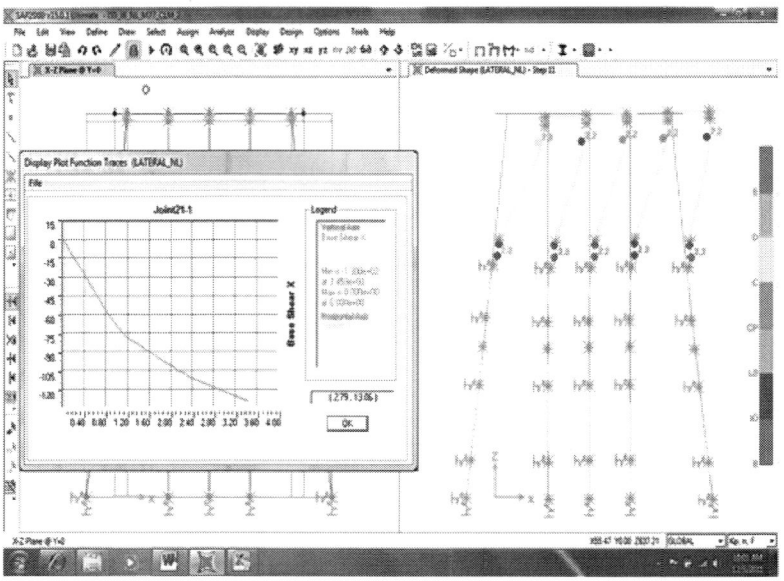

Figure 17: Plots of pushover response to lateral load at bearing positions of intermediate bent pile cap for one of the Coldwater River bridges studied [16] (see Figs. 5 and 10); left plot shows force-displacement response; right plot shows deformed shape and limit state condition for plastic hinge locations.

In assessing the results of the time history analysis, a simplified analysis was performed using a single DOF equivalent model using a lumped stiffness based on a unit force lateral load analysis. When compared to the linear dynamic analysis of the full bent, this simplified analysis was able to demonstrate that the first mode of the system dominated the overall response and could have been used as a predictive tool.

A decision was made not to depend on nonlinear dynamic response analysis for all the bridges. This was in part due to the lack of confidence in the soil properties at the site from which reasonable assumptions could be made for simulation of nonlinear soil response and in part due to the scope of the work which was limited as has been mentioned previously. A limited attempt was made to verify at least the elastic properties of the soil adjacent to the sites using seismic refraction tests performed near the embankments which were accessible on dry land.

CONCLUSIONS

This chapter highlights objectives, modeling options, and analysis results for a range of FE based vulnerability assessments of highway bridges performed at the University of Mississippi by the author. To illustrate the range of conditions and considerations, three projects have been selected as case studies. Finite element software, and the operating systems and hardware, especially microprocessors which support the software, have advanced significantly over the time period since the first of these projects was conducted. The lessons learned, however, remain fundamental in a sense for vulnerability assessments that are premised on mechanics of highway bridge materials, elements, and structural systems that inevitably include soil and construction materials such as concrete and steel.

The objectives of the vulnerability analysis depend in part on the nature of the hazard, the inventory exposed to the hazard, and the agency concerned with the inventory. For the study cases the hazard for the region is characterized by a high consequence but low probability event. The inventory is not near enough to the hazard to be considered in a high seismic exposure but the potential ground shaking is significant enough to inflict severe damage especially to older bridges that pre-date seismic design provisions. Furthermore, and perhaps of most interest to the state emergency management and transportation officials that have sponsored the studies, the bridges selected for detailed evaluation are located on important lifelines between a major metropolitan area and the multiple surrounding communities that are growing rapidly. Many of these communities would become isolated in the event of the complete functional loss of the highway network. Both urban and rural stakeholders will depend on these bridges remaining serviceable not only for the densely populated area closer to the hazard needing to exit the concentrated region of potential seismic damage but also for incoming emergency responders and other personnel providing assistance. The understanding of both the hazard and the inventory in the study region is evolving even at the present time, and a research study is now underway. The study is exploring the short and long term impacts of potential damages on traffic flow in north Mississippi as well as the resulting economic losses. The development of seismic ground motion records for the study cases is addressed only to the extent necessary to characterize the hazard and help interpret the results obtained from the FE simulation. In most situations the input motion is a major uncertainty in both the model analysis and the vulnerability assessment. A performance based approach has been adopted where consistent with the study objectives. A range of hazard and ground motion intensities has been considered in these studies and FE based time history response analysis has formed the basis of the performance evaluation. In one case, the objective was to validate results of a regional study that did not consider many of the key details of the structural system using FE based analysis. In this case, the hazard and ground motion intensity were selected to be consistent with that used in the regional study.

Examples of 2D and 3D structural models are presented that incorporate a wide variety of finite element and material types and consider the effects of soil-foundation-structure interaction which, in the view of the author, is an essential part of reliably establishing the performance of bridge structural systems. The incorporation of such interaction presents challenges to the analyst which are not well represented in standard textbooks on highway bridge design and even some of the FE literature. To complement such reference works, a portion of the chapter is devoted to discussion of preliminary analyses that are quite naturally performed while the more complex models for final evaluation are constructed. The discussion highlights the importance of first capturing behavioral aspects of the system revealed by static response analysis under gravity or idealized lateral loads and subsequent examination of vibration mode shapes and natural frequencies obtained by eigenvalue analysis. These preliminary analyses provide not only quality assurance but also insight that may guide expectations for the results of the more complex models. The information obtained from subsequent static, nonlinear, and dynamic response analysis is then maximized so that the most useful or telling information is extracted from the analysis under seismic excitation.

The FE based approach to vulnerability assessment ensures that quantitative data formulated on basic mechanics principles is generated for consideration during the assessment. Extracting the data and using it to establish measures of performance remains somewhat of an art. In the study cases, a range of measures has been adopted including peak dynamic response acceleration and displacement as well as maximum internal forces in critical members and damage distribution in major subsystems. It is hoped that an appreciation of the complexity of highway bridge systems has been provided through the description of the many details of the FE models and the results obtained from analysis of response to seismic excitation.

Application of the results of FE analysis to a specific vulnerability assessment requires consideration of the objectives and end-user needs. A range of complexity in successive models used in the evaluation may be appropriate depending on the sensitivity of the evaluation on the outcomes of the analysis. Furthermore, the availability of powerful analysis tools should not overshadow lack of confidence in data provided to the analysis. In particular, soil property and earthquake intensity and motion characteristics are often not known precisely.

In regions of moderate seismic hazard it may prove difficult to establish a sense of urgency for action on the basis of the results of a vulnerability analysis whether or not is based on FE modeling and considered highly accurate. In such a context it may be useful to incorporate the seismic vulnerability assessment in a broader one considering multiple hazards exhibiting comparable levels of risk.

ACKNOWLEDGEMENTS

The material in this chapter represents the accumulated effort of the author with others involved in the cited bridge seismic vulnerability studies. These include Dr. Robert Hackett, former chair of the Department of Civil Engineering at the University of Mississippi, and Mr. Charles Swann, now Associate Director of the Mississippi Mineral Resources Institute. I would also like to thank the many graduate research assistants in the Department of Civil Engineering at the University of Mississippi under my advisement who contributed to these studies and used them as the basis of master's theses and a doctoral dissertation. The academic works are cited separately in the relevant references given. Lastly, the financial support of the Department of Civil Engineering is appreciated as is that of the various sponsors of the cited studies. Sources of the latter include two FEMA mitigation grants administered by the Mitigation Bureau of the Mississippi Emergency Management Agency and a federal grant administered by the Research and Bridge Divisions of the Mississippi Department of Transportation. The specific sponsored project and grant details are described in more detail in the relevant references given

REFERENCES

1.   Computers and Structures, Inc. SAP2000 Users Manual-Version 15. CSI; 2012.

2.   Computers and Structures, Inc. CSIBridge Users Manual-Version 15. CSI; 2012.

3.   Elnashai AS, Jefferson, T, Fiedrich F, Cleveland LJ, Gress, T. Impact of New Madrid Seismic Zone Earthquakes on the Central USA:- Volume I. MAE Center Report No. 09-03. Mid-America Earthquake Center, University of Illinois; 2009.

4.   Federal Emergency Management Agency. FEMA 273: NEHRP Guidelines for the Seismic Rehabilitation of Buildings. FEMA; 1997.

5.   Federal Emergency Management Agency. HAZUS-MH (Hazards US-Multihazard)-Version MR5. FEMA; 2010.

6.   Federal Highway Administration, United States Department of Transportation. National Bridge Inventory. http://www.fhwa.dot.gov/bridge/nbi.htm

7.   Hibbitt, Karlsson, Sorenson, Inc. ABAQUS Theory Manual-Version 5.6. HKS; 1996.

8.   Hwang H, Huo JR. Attenuation Relations Of Ground Motions for Rock and Soil Sites in Eastern United States. Soil Dynamics and Earthquake Engineering 1997; 16: 363-372.

9.   Ismail IMK, Mullen CM. Soil Structure Interaction Issues for Three Dimensional Computational Simulations of Nonlinear Seismic Response. EM2000: proceedings of the 14th Engineering Mechanics Conference, 21-24 May 2000, Austin, TX. Reston, VA: ASCE; 2000.

10.   Johnston AC, Schweig ES. The Enigma of the New Madrid earthquakes of 1811–1812. Annual Review of Earth and Planetary Sciences 1996; 24: 339-384.

11.   LeBlanc B, Mullen CL. Characterization of Abutment-Deck

Interaction using 3D FEM and Field Vibration Measurements for an Existing Highway Bridge in North Mississippi. In: Uddin, W, Fortes RM, Merighi JV. (eds.) Proceedings of the 2nd International Symposium on Maintenance and Rehabilitation of Pavements and Technological Control, 29 July-1 August 2001, Auburn AL. NCAT; 2001.

12. Mullen C L, Swann CT. Seismic Response Interaction between Subsurface Geology and Selected Facilities at the University of Mississippi. Engineering Geology 2001; 62(1-3) 223-250.

13. Mullen C L, Swann CT. The State of Mississippi Standard Mitigation Plan-Earthquake Risk Assessment. Final report to Mississippi Emergency Management Agency-Mitigation Division. Center for Community Earthquake Preparedness, University of Mississippi; 2004.

14. Mullen CL. Seismic Vulnerability of Existing Highway Bridge Substructures Supporting the I-5 Undercrossing at MS302 (Goodman Road). Final report to Mississippi Department of Transportation-Bridge Division. Center for Community Earthquake Preparedness, University of Mississippi; 2001.

15. Mullen CL. 3D FEM for Seismic Damage in a Four-Span Interstate Concrete Highway Under-Crossing including Embankment-Structure Interaction. Proceedings of the FHWA National Workshop on Innovative Applications of Finite Element Modeling in Highway Structures, 20-21 August 2003, New York City, NY. UTRC; 2003.

16. Mullen CL. Seismic Vulnerability of Critical Bridges in North Mississippi. Final report to Mississippi Emergency Management Agency- Mitigation Division. Center for Community Earthquake Preparedness, University of Mississippi; 2011

17. Swann CT, Mullen CL, Hackett RM, Stewart RK, Lutken CB. Evaluation of Earthquake Effects on Selected Structures and Utilities at the University of Mississippi- A Mitigation Model for Universities. Final report to Mississippi Emergency Management Agency. Department of Civil Engineering, University of Mississippi, and Mississippi Minerals Resources Institute; 1999.

18. Tuttle MP, Schweig ES. Archeological and Pedological Evidence for Large Prehistoric Earthquakes in the New Madrid Seismic Zone, Central United States. Geology 1995; 23(3) 253-256.

19.   United States Geological Survey. Earthquake Hazard in the Heart of the Homeland, Fact Sheet 2006–3125. http://pubs.usgs.gov/fs/2006/3125/pdf/FS06-3125_508.pdf

Chapter 3

Model for Geologic Risk Management in the Building and Infrastructure Processes

Liber Galban Rodríguez[1]

[1]Universidad de Oriente, Constructions Faculty, Hydraulic Engineering Department, Cuba

ABSTRACT

The geologic risks management is a process that requires to follow the tendencies of the new models of technological innovation. Nowadays it becomes necessary to elaborate an specific model for the management of the geologic risks, that is adapted to the peculiarities of the current development of the building and infrastructure systems; and allow the use of the current tools as the GIS, Wombs, Analysis Cost

Benefit, etc., for the organization and the control of the knowledge management and final quality of the executed works. To model with the processes management could be an alternative form before this task. Proposing in this occasion a variant to negotiate from this perspective the management of geologic risks in the building and infrastructure processes.

INTRODUCTION

According to the resulting comprehensive geological science, many scientists in other fields tend to erroneously point to some primary or secondary geological events as not owned by or for study by geologists. This interpretation of the insufficient knowledge of geology as a science, mother of geosciences, and the fields and branches of this science. A summary of the sources suggests that, in principle, the geology is the science that studies the formation and origin of the Earth and its component materials inside and out, as well as, the study of all phenomena and physical and chemical processes natural, and its evolution over time, taking place on the planet Earth from its own emergence, focusing greater focus to those that occur in its outer part, or the crust.

Understand then, for example, the relationship between atmospheric phenomena and their impact on the earth's crust are studied by this science, or that relations between phenomena that originate within the earth with clear consequences in climate and our atmosphere, are also studied by geology, is a logical question for geologists. So also the actions performed by men and affecting one or more components of the earth's crust and the evolution of terrestrial flora and fauna and their footprints on the rocks, are also under consideration, among others, science geology.

Important aspects of this science are the geological processes and phenomena, also known geological events. The geological events taking place on planet Earth, and create transformations that occur in a slow or sudden. However, each may be equally fatal to society depending on a number of factors that are discussed below.

The planets own forces are born of the Earth, but project their effects in different ways in the land surface and the outer space. These forces include gravity, magnetism, physical-chemical reactions and geological processes associated with them. Taken together generate the tectonic plate movements, surveying and land decreases, the eruptions of volcanoes, geysers and fumaroles, springs, earthquakes, tsunamis, changes in relief, the secular changes of climate and a varied range of events related to the formation and transformation of substances and the landscape. In summary, internal forces of the planet determines the landscape of the earth's surface, whose influences on the environment and life are crucial for the present and the future of society (Iturralde-Vinent, et al, 2006).

Slow or cumulative events are those that act over a long period of time, so that its effects are evident by inspection. The assignment to the environment and society of these events occurs through the accumulation, in addition, tens of thousands years. For example, karst processes, where cavitation's occurs and subterranean (popularly known as "caves"), changes in the relief surface (hummocks, among other forms) (Figure 1), or the presence of small concentrations of substances harmful in rocks, soils and natural waters, which were not detected by specific studies, and they can concentrate to unhealthy levels due to the consumption of water and plant to be drawn from these media.

Figure 1: Karst formations, wooded hills of the Viñales Valey Pinar del Rio, Cuba. Photos: grind León, 2004,http://www.mappinginteractivo.com/plantil-lante.asp?id_articulo=815.

Other events are slow secular movements of the ground, which typically occur at speeds that are measured in millimeters per year, but eventually come to cause major changes in the topography and buildings affect the coast, or over the rivers. By contrast, sudden event, usually catastrophic, are those that occur by the release in a short space of time, some energy inside the Earth and its combination with external phenomena, resulting in volcanoes, earthquakes (Figure 2), landslides, mudslides, floods, etc. (Iturralde-Vinent, et al, 2006).

Figure 2: Sudden geological event. Earthquake Haiti, registered on January 12, 2010 at 16:53:09 local time (21:53:09 UTC) with epicenter at 15 km from Port au Prince, Haiti's capital. Views of National Palace and collapsed buildings in downtown Port au Prince.

Hence, to know what kind of events can occur in the future in a given region, although not known exactly when and at what level can occur, is an activity of fundamental importance in guiding the development of a region, so that the impact of these events is the minimum possible and do not pose a disruption to the social and economic development of it. Knowing the potential effects and / or losses that may occur in the social and material allows within development plans and investment programs, you can define measures to prevent or mitigate the consequences of future disasters, whether through involvement in the occurrence of the event, if this is possible, or modifying the conditions conducive to its effects occur.

GEOLOGICAL RISKS

Geological risks are part of a broad set of risks that would be encompassed between environmental hazards, and grouped into classes according to their origin. The definition of geological risk has been addressed by several authors. One of its early definitions, formulated by the U.S. Geological Survey in 1977, states that geological risk means any geological condition, process or event which represents a potential threat to the health, safety or welfare of a group of citizens or functions of a community or economy. Geological risks cannot arise from simple description of the material or natural processes. Not conceive, either, regardless of the purpose for which they can cause on people, on their work or in general on the ecological balance (Brusi, 2003).

According to Ayala (1992), geological hazards are those processes, events or situations that take place in the geological environment and can cause damage or harm to communities or infrastructure that are vulnerable zones occupying a territory. Also understood as a process, situation or event in the geological, natural, induced or a mix that can generate economic or social harm to any community, and whose prediction, prevention or correction geological criteria are to be employed. Another definition are understood as a circumstance or situation of danger, loss or damage, social and economic, due to geological condition or a possibility of occurrence of geological process, induced or not. (Ogura - Macedo Soares, 2005). It is also distinguished, which are defined as processes occurring within the sediment (building, gas generation, break-cementing,...) and require no action by external actors and those who are conditioned by the action of some external factor, natural (volcanism, uplift, subsidence, tectonic collapse, diapirism, currents, tsunamis, hurricanes...) or artificial (fluid extraction-gas-or oil, etc).

They all agree that geological hazards can be caused by natural or induced. In this sense, there are situations in which man's interaction with the environment that creates a potential risk situation, since human action itself has a "trigger" mechanisms to natural hazards or natural geological events could pose a or generate social harm and / or economic (Orberá - Ramirez, 1994). Geologic events that could represent potential threats to society, characterized by its unpredictability and its deadly consequences, but more dangerous is the degree of ignorance that exists at various levels on the types of risks they generate. Several authors have worked on the lines of classification of geological hazards, most of them agree classified according to the conditions that gave rise to them, namely:

- Natural geological risks
- Geotechnical risks.

Geological risks of natural kinds are those that are not produced at source by the hand of man, although could empower, they can originate from inside the Earth because its structure and together are known as endogenous or come from outside and are called exogenous. A summary of the literature describes them according to exogenous or endogenous origin is as follows (Galban, 2009): O is for errors of calculation and estimation of physical - mechanical properties of the soil, the failure of natural geological processes and phenomena and to non-works adaptation of certain parameters of resistivity, with the actual probability of occurrence of disastrous events natural or technological. And those caused by population growth, intensive agriculture in unsuitable areas, lack of evaluation of different types of long-term effects, etc. (Galbán, 2009).

Table 1: The geotechnical risks are induced geological hazards and enhanced by human error of calculation and lack of prevention in civil engineering

Endogenous Geologic risks	Earthquakes, volcanic eruptions, liquefaction or liquefaction, tectonic movements, Tsunamis, karst , natural gas and hazardous substances, hydrothermal mineralization, cracks, cavities and landslides collapses, expansive soils, land subsidence
Exogenous geological risks	Storms, hail, cyclones, tornadoes, coastal flooding, river flooding, overflows of rivers and streams, erosion and sedimentation, impact of meteorites, salinization, desertification and drought, wind erosion, landslides, rockslides, avalanches

Too many examples of risks induced by human activity, some examples include: landslides resulting from the change in the balance pending the construction of roads, broken dams or reservoirs (Figure

3), the subsidence of the land by mining, overuse of aquifers or tubing associated with water pipes, earthquakes triggered in rapid filling of reservoirs, settlement, subsidence and cracks of buildings on soft ground, among others.

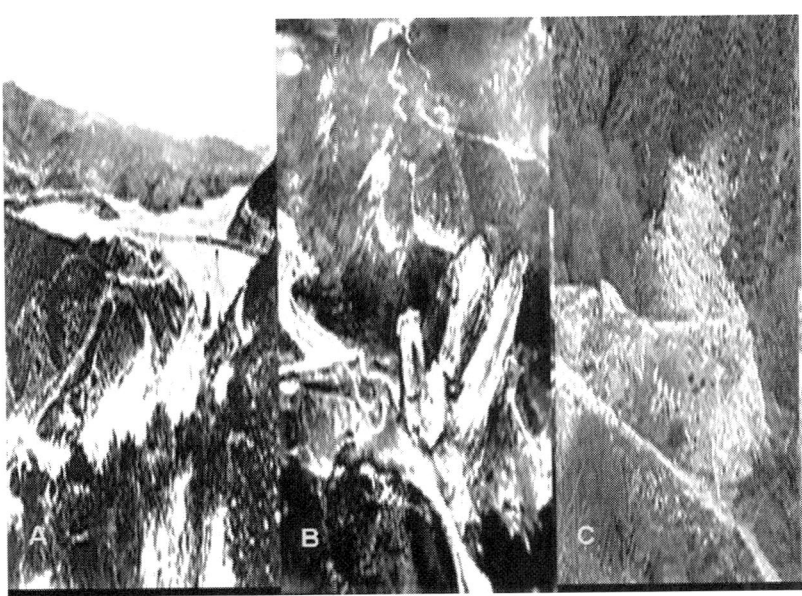

Figure 3: Saint Dam Disaster. Francis, Francis, Los Angeles County, California, USA. Completed in 1926, the March 12, 1928, catastrophically failed due to geotechnical calculation errors during execution, killing more than 600 people. _Francis_ Dam Images from the start of the gap (A), after the disaster (B) and current image of the remains of the dam base (C).Http://en.wikipedia. org/wiki/St._Francis_Dam.

The different types of geological hazards can interact with each other, and in the present predominance of one other side effect, which can complicate the situation and increase the vulnerability of the object of work in question. Because we cannot conceive without independent analysis finally perform a risk assessment as a system, supplementing these with geophysical, geodynamic, geomorphological and hydrogeological risk maps, etc.), Which in the literature does not appear specified in this way, although if certain risks related to or associated primary and secondary or used geographic information systems to determine a certain level of risk.

These questions denote that the geological risk in terms of construction and infrastructure projects, whether it is characterized, it is also necessary that depending on the use of this knowledge, take administrative measures and technological lead to ensure a certain level of safety therein.

THE GEOLOGICAL RISK MANAGEMENT IN THE BUILDING AND INFRASTRUCTURE PROCESSES

Management is a modern concept, an issue that brings together aspects such as research, planning, organization, evaluation, management, analysis, implementation, monitoring and control (Kootz, 1998). Meaning that, properly inserted according to mitigate geological hazards, is a very useful working tool in the construction processes and infrastructure.

Considering all the prerogatives analyzed, taking into account the concepts related to the previously defined geological risk is defined for this investigation and management of geological risk, the activity which is responsible for the studies to be made of the phenomena or processes related to land and geodynamic processes or phenomena induced by human activity that affect projects and / or works of engineering, civil infrastructure, situated or in the future be located on the ground, so that these help plan, organize, manage, evaluate and control the organizational measures, techniques or technology that are issued for these projects or works, aimed at preventing or mitigating the effects of disasters caused by geological events of natural or anthropogenic (Galbán, 2009).

More broadly we can say that the geological risk management is performed to predict the consequences (risk) that future geological phenomena and natural or induced processes (risk) will have on a particular work or project which conceived man takes implicit or no transformation of reality (vulnerability) and therefore it becomes necessary to make organizational and technological measures to reduce its impact (management). (Galbán, 2009)

The biggest problem is that risk management is a problem internationally long term, decision makers have not always been particularly good at planning long-term development, or have spent much money in reducing these long-term risks (Monge, 2003). Therefore, precisely because their role is aimed at carrying out certain transformations of reality, needs to be contextualized and based on this pose a mechanism enabling the extent of the real possibilities of each country.

The risk may generate an infrastructure construction project and may be permanent or recurrent, affecting the daily lives of people and possibilities for development of an area or region in general. Also a risk that translates into a disaster, the event must be of a very large, as in some cases a series of small events, caused or enhanced by the construction of an infrastructure project may be more disastrous one of considerable magnitude. Similarly, a small phenomenon may be a warning that conditions are brewing risk in the future, may lead to a disaster of great magnitude.

The effect of construction and infrastructure projects in the generation of risk can occur in two ways: In the process of construction and operation, when trigger reactions of nature such as floods, droughts and landslides, especially when they cause deforestation, Inadequate management of soil, drainage and flood areas, wetlands, or artificial fillers between some elements. And the other way to generate risk is due to the permanent exhibition of the construction projects and infrastructure to natural geological phenomena induced which multiplies the effects on people and ecosystems in general (Monge, 2003).

To reduce the risk in the construction processes and infrastructure can be put in place, both prevention and mitigation, so that the effect is minimal. The prevention is to avoid or prevent natural events or generated by human activity are causing disasters. For its part, mitigation is the result of an intervention designed to reduce risks, trying to change the nature of the threats, in order to reduce vulnerability, so that it would mitigate the potential damage on the life and property (Cardona, 2001).

Correspondingly, one should consider that any measures designed to reduce or eliminate a risk, is closely related to processes in the medium and long term established for the development of a country or region, why should be incorporated into programs upgrading of enterprises implementing construction projects, or what is the same, should be incorporated into a management process, a process that should be developed or designing using different measures or tools. Today, these measures fall into two basic types:

- Structural measures.
- Non-structural measures.

Structural measures of prevention and mitigation are employed engineering works to reduce or lead to "acceptable" levels the risk that a community is exposed. They run directly on site and can be classified as preventive or corrective control. Its construction requires engineering design and optimization of resources, as well as, an Environmental Management Plan that will enable the reduction of the impact generated (Collective of authors. 2005).

There are several types of structural measures for treatment of landslides, erosion, floods, torrential floods, earthquake damage, among others, some of them are:

For landslides: The removal and/or shaping the contours of the ground or slope, which is performed in order to increase its stability, an issue that can be achieved by building trenches stabilizers, shares of terracing, coated plants or artificial among others.

For river erosion is primarily used coating with mulch, waterways, in filtration trenches, among others. For flood expansion works are performed or misuse of causes of rivers, building dikes and dams, etc. For earthquakes, for example, structural reinforcements are made in buildings by applying methods of geometric configuration, such as the static equivalent method and the modal analysis method, combinations of shapes are made, certain factors are calculated using both the depth and the area of foundations and reinforcements that are necessary to implement these, including specifications for embankments, slopes and near buildings, among others.[1]

These measures will positively impact the environment, quality of life of people living in areas at risk and during the construction phase generate employment. However, they can affect the health of the population, the lifestyle of the community and the mobility of pedestrians and users, and can generate negative impacts on different environmental components in each phase of construction of the project, therefore requires the implementation of actions to minimize these impacts (Collective of authors. 2005).

One way to force developers to implement certain structural measures during the execution of works, is through the adoption of codes or construction standards. In most countries, were adopted in various standards or codes that in one way or another to geological risk management processes and infrastructure construction, within these processes and focused on building and infrastructure, meet the standards for earthquake resistant construction, the project documentation, execution of works, geotechnical standards, among others. These rules indicate what calculations during the execution should be performed, how they should implement certain measures, among other things.

Non-structural measures are the most simple and important, and the most used around the world since ancient times. These bring together a set of functional elements related to physical planning and land use, technological tools, education, observation, legal, administrative, among others, which also help manage geohazards indirectly, within which include:

- The design of models, methodologies, strategies, software, among others, to study, assess, manage..., management of geological risks.
- The planning of land use, and with this construction that they are running.
- The legislation of environmental factors that influence the management of risks.
- The incorporation of preventive aspects of the budgets of state and private investment.
- The organization of national and international scientific networks techniques for the investigation of the behavior of different events and associated risks, as well as project development and exchange of experiences.
- The organization of monitoring systems and early warning.
- Other specific measures depending on the types of risks.

There are other methods as those used in the assessment of environmental impacts, such as checklists, matrices, networks, cost / effectiveness / benefit and multi-dimensional models, which could be adapted to estimate the risk (Clarke, 2001) also providing rigor and accuracy requirements needed in the construction processes and infrastructure.

Besides this, it is always necessary to deepen local knowledge, timely, necessary dig into the specifics of each region, and that includes climate, geology, anthropomorphism, history, population characteristics, intent of use, etc., Or for the management of geological risk, one must also have completed certain steps of knowledge acquisition, both in individuals who perform the management and the institutions responsible for the investment (Galbo, 2009), all in an environment of multidisciplinary.

A current variant is the adoption of models. A model is the result of the process of generating an abstract representation, conceptual, graphic or visual phenomena, systems or processes to analyze, describe, explain and simulate these phenomena or processes. [2] Today's systems or models of technological innovation are becoming increasingly complex. The assimilation of new technologies is not a passive, nor is achieved only by training the technical staff and operators in other countries as often happen. They need a culture around these technologies, an entire local culture in which staff training is based on domain knowledge and in depth, the laws and principles that govern it. This allows not only operates efficiently, but face new and unexpected situations, make necessary adjustments and innovations creatively develop increased on the same (Group of authors. 1999).

On the other hand, it is known that many scientific results in terms of disaster risk management are not applied in business practice, in many cases, issues with economic and institutional factors, characteristic of the international situation and other by administrative status, knowledge, organization, control management (Galbán, 2009). This is compounded by the low disclosure in the world of the results obtained by many scientists for its widespread use, the virtual absence of focal points, and the need to develop an awareness and appropriate calculations as to the levels existing geologic hazards and risks.

PROCESSES MANAGEMENT AND GEOLOGICAL RISKS MANAGEMENT

A late of the eighties of last century, and derived from the need to increase the quality of economic and productive processes of enterprises in the developed capitalist world, there is a new management tool, which initially was called or process management process approach, this tool, in the year 1994 was adopted by the ISO as a standard for improving quality management, ISO 9001. Since its emergence has had several subsequent versions in 1998, 2000, 2001, 2003 and most recently in 2008.

Process management can be conceptualized as how to manage the entire organization based on the processes, these being defined as a sequence of activities to create added value on an entry to get a result and an output which in turn satisfies customer requirements (Negrin, 2006).

The process approach is based on:

- The structuring of the organization based on customer-facing processes.
- The change of the organizational structure from hierarchical to flat.
- Functional departments lose their raison d'etre and are multidisciplinary groups working on the process.
- Managers and supervisors fail to act and behave like cowards.
- Employees focus more on the needs of their customers and less on standards set by his boss.
- Using technology to eliminate activities that do not add value.

The process approach requires a logistical support, which enables the management of the organization from the study of the flow of materials and associated information flow from suppliers to customers. The customer orientation, or provide the service or product for a given level of satisfaction of the needs and requirements of customers, represents the fundamental gauge of corporate profits, thus obtaining an efficient supply management and timely response to the planning process.[3] Companies and organizations are as efficient as are their processes, most of which have become aware of what was previously stated, have reacted to the inefficiency representing departmental organizations, with their niches of power and excessive inertia to change, promoting the concept of the process with a common focus and working with an objective view on the client. [4] The main advantages of this approach are:

- Align organizational objectives with the expectations and needs of customers
- Shows how to create value in the organization and
- Points out how they are structured flows of information and materials
- Indicates how actually does the work and how to articulate the customer supplier relationships between functions.

The process approach is currently applied in conjunction with the theory Denim Cycle [5] which in principle suggests that the quality management processes generated by an activity must be cyclical and is in line with four stages: Plan, Do, Check and act. This means that an organization should always be improving corporate acting or correcting previously planned and done to improve it or what is the same as continually improving the management of the company, also allowing the products or services in the process of exploitation and consumption, become real laboratories that process.

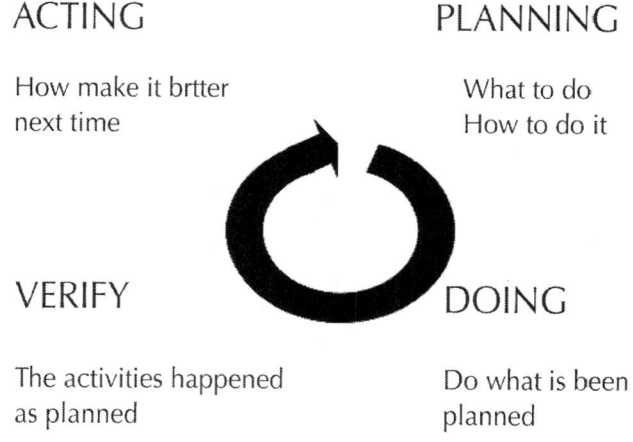

ACTING

How make it brtter
next time

PLANNING

What to do
How to do it

VERIFY

The activities happened
as planned

DOING

Do what is been
planned

Figure 4: Denim cycle.

For the implementation of process management approach to an organization, it is essential among other things, create the necessary cognitive and technological conditions. Many companies take years to implement it in its entirety, and its implementation, first requires a thorough investigation of the behavior of all components of the organization in all its facets, or must do science. It also requires a strategy in the medium and long term. The most common is to be introduced in stages or subsystems, for example, sub-economic management, human resources, design, general services, production, etc. Attached to this is to identify an approach is also used certification of compliance with its requirements. This certification is done internationally by the ISO, which assigns a panel of arbitrators or advisers, who are responsible in different countries to carry out the audit inspection process and, finally, after verifying in practice correspondence, from the extension of the certificate of quality compliance with ISO 9001 in the subsystem inspected. This certificate has an important significance, as it proves to other organizations or outside this sector, and society in general, the activity, product or service they perform, comply with all requirements necessary for the purpose with which designed and with high quality, that also increase the prestige of the organization to the international community.

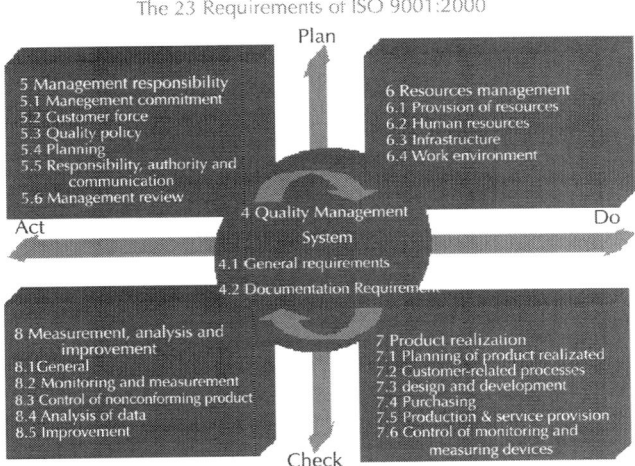

Figure 5: Requirements of ISO 9001/2000.

It should be noted that under the principle of managing processes in the world have been many working tools in various areas of human development, so much so that several of the ISO standards that emerged later, are also developed in the environment processes.

The current management of construction projects, regardless of their particular characteristics, is moving steadily towards process-based schemes, such as in the rest of the industry and services. These processes are not always well defined, lie necessarily in the implementation of quality systems and its far more classical definition (quality control, quality assurance) and involving the full set of activities to be developed. However, for the client of a construction project, there are certain processes that are more significant, in that they affect their own effectiveness as a manager, than others, which nevertheless still important in the entire business.

Perhaps the three most significant groups of processes for the customer are those relating to the economic control of the project (quantitative control), those that affect the quality of the product will receive (quality control) and, finally, the fulfillment of milestones in execution (control limits). The processes listed above, are supported by others who have most influence on those who carry out the project, such as the administration itself, the implementation of the various fractions of the project, etc.

The whole process generates a significant amount of documentation that must be preserved, distributed and evaluated. Contrary to the widespread view, this documentation should not have a volume greater than if quality systems are applied to production.

Transfers of technology in business management and management of geological risks to the developing countries, suggest the analysis of the technological, environmental conditions, social and economic conditions of each country. The advantage of representing the process management technology improves several aspects of business management, where the management of quality in their services or products is increased and enhanced in particular. International experience has acknowledged progress developer in the implementation of process management in various facets of economic and social development of countries, is considered a relatively young subject and novelty, which calls for more research to accurately set and increase aspects in the ISO standards, which do not include the management of geological hazards within their applications. The management of geological risks is also a process that has certain peculiarities in civil engineering projects or hydraulic. If you need to understand the process approach, using its bases to the management of geological risks in these projects is developed, it is necessary interpret its components, such as "organization" would be the construction company executing the project construction or infrastructure the "customer" would be the investor, the "processes" are the stages of the project, the "threads" could be for example the seismic risk assessment in the preliminary stage, and the procedure could be the way to proceed with the seismic risk assessment. There is no difficulty in the interpretation and application of general principles of process approach to the management of geological risk in the construction processes and infrastructure, an issue that also pursue the same objectives of the approach and its advantages.

Taking into account that eventually the management of geological risks in construction and infrastructure works, which is looking to improve the final quality of them to be better able to withstand the geological events. You can say then that is correct adopt the principles of process management to model the geological risk management processes and infrastructure construction companies and develop institutions for them. In other words, this management can be implemented within a technological paradigm based on process management.

MODEL PROPOSITION FOR GEOLOGICAL RISK MANAGEMENT

The knowledge management model proposed in this contribution, part of the recognition of the need to improve the management of geological hazards in the construction processes and infrastructure, made by individuals and institutions directly or indirectly involved in them, and used for this description of the steps or actions in the threads that make diagnosis, design, implementation and evaluation. Its aim is to show the functionality of the indicators analyzed in stages or diagnostic procedures, design, implementation and evaluation, which can be developed to express and evaluate the organizational management of geological risk. Moreover, this modeling is not inconsistent with the desires and objectives of the regulations in force internationally, the problem is that according to the analysis performed, there is no single technology model that meets the necessary requirements, enabling approved and unifying quality criteria as far as geological risk management concerns, and also follow international standard patterns for these issues since the project is conceived until its conclusion. The risk management model proposed geological, is functional at the same time, and is a representation of what could be an alternative and inclusive knowledge management, which serves both the organization and its environment.

The proposition of the model is based on different aspects that must be met, and which form part of the international situation discussed above, these include:

- The investment process.
- The system of codes, rules and current legal regulations, which intervene in the management of geological risks.
- Processes management.
- The reality of the construction companies.
- The measures, regulations and national and local policies, proposed and implemented by the government and institutions.
- The international conventions and treaties on environment and disaster management.
- Multidisciplinary involvement in research and implementation of solutions.

The tasks to be carried in every action of the processes are subject to the conditions to be created in each organization and can be used various procedures and techniques such as Benchmarking, Reengineering, the SWOT matrix, among others.

Table 2: General components of the model

Processes	Actions
Diagnosis	Analysis of the current situation.
	Establish working definitions.
	Establish current strategic position.
	Analysis of resources.
	Requirements Analysis.
Design	Development of strategy knowledge.
	Definition of strategic goal.
	Architectural design knowledge.
	Creating organizational climate.
Implementation	Implementation of the plans developed.
	Revision of the strategy.
Evaluation	Implementation of measurements.
	Interpretation of results.

The resulting model of our research should follow the steps raised in a general methodology designed for different stages of project implementation, by adding to these the one conceived by Denim, or continuous improvement. So this methodology includes four stages of geological risk management, for whom and under what is deducted from the literature review, described then what are the key actions to be performed.

- Preliminary Stage (diagnosis and design processes)
- During the project implementation. (Part initial implementation)
- Stage of project completion. (End of implementation)
- Continuous improvement process. (Evaluation Process)

PRELIMINARY STAGE

At this stage the companies and institutions conceived the basic ideas of the project, perform diagnostics, designs, application for licenses, permits, contracts, literature review, etc. For the purposes of the model from two of its components:

Diagnosis Process

The aim of the diagnostic process is to determine the corporate resources that express the knowledge of the organization and its use to propose projects that allow the representation of organizational knowledge, their development and use in the qualitative improvement of the organization. The actions included in this general process are:

- **Current Situation Analysis**: The diagnosis is performed to know the current situation, the result of the completion of this process is to guide the action plans within the strategic development of the organization.

- **Establish Working Definitions**: It is necessary to establish a working definition of what each organization means knowledge. For an entity, can be "patent", in other capacities or also "experience". In our object of analysis, states that knowledge is reflected in documents, methodologies, procedures, reports, maps, etc. On the other hand has to do also with the participation of specialists from different disciplines, both in the pursuit of knowledge and the training of trainers.

- **Set the Current Strategic Position**: It means identifying the level of access or knowledge barriers. This analysis provides the following categories: special, temporal and social. That is, where they reside (entities), what is the relevant time-frame of organizational memory, knowledge sharing, among others, and what is the hierarchical, functional and cultural context is contextualized, that which impedes or promotes the exchange of knowledge.

- **Resource Analysis**: seeks to identify the categories of knowledge that exist, requires the identification of internal and external sources, such as research and development, relationships with other entities, sources that exist or are used in the organization, their relationships, the level which is currently and the level to be achieved.

- **Requirements Analysis**: Understand the requirements associated with implementing the project, analyzes the nature and the project environment, functionality and action plans.

At this stage, proposed to the specific management of geological risk carrying out the following:

To determine the social use of the work and general characteristics.

- Make a diagnosis, which take into account the most relevant research results, the available historical information on the occurrence of significant events in certain localities with the resulting effects, or that is available in the institutional archives, the analysis of the difficulties in place to deal with a real natural phenomenon caused by man or the combination of these, besides all that useful information that could be taxed at a better management of geological risk (information management)

- Analysis of the information provided by geological and engineering geological reports earlier reports from the study area or nearby, enabling management geological risk.

- Analysis of data and information provided by the Geographic Information Systems.

- Study of the surrounding environment, identification of key activities related to social and business discipline. (Socio-environmental risk management).

- Study watersheds (surface and groundwater), their relationship to the threat of occurrence of severe weather events and the environment. Influence in the region of study. (Hydro geological risk management).

- Selection of appropriate methods or techniques to analyze the information obtained.

- Interpretation of the relationship between the occurrence of various natural and human phenomena possible to present the proposed work, which should lead to knowledge of the potential presence of danger and the behavior of the levels of vulnerability of areas of investment. Or more broadly stated, total identify threats, vulnerabilities and risks, identify possible single or combined (systematization of geological risk management).

- Identification and review of the main rules governing the implementation of these activities.

- Fabrication of the chips in the process, explained the contents and tasks of each thread of the model for this stage aimed at reducing vulnerability constructive.

The basis on which rests the whole structure of the integrated management of a construction, is the uniform treatment of information and capacity building of knowledge. This also means, uniformity in the processing of documentation, regardless of its source, its origin and its subsequent use (Serra - Pérez, 2007), the implementation of field investigations by specialists in preparing for interviews, surveys, assessments quantitative and qualitative economic, among other techniques, as well as in the training of technicians in areas related to process management.

The ease of use of databases and spreadsheets trade has meant that much information is treated by more and more people within the organization. However, well-managed construction organizations, tools for analyzing data sets are, with few exceptions, non-existent. It is rare to find tools to cross, for example, production data with quality, and even more difficult to analyze in some other way such data relationships. For these reasons should be narrow as well, which will be or what techniques or methods used to collect and analyze information, and what are the specialists who participate in this discussion, always valuing multidisciplinarity.

In the geological branch in the world are already being implemented tools such as databases that may well be used for risk management, which allow you to organize, process, transform and transmit information to the territory in question, quantitative data and formats, qualitative, logical and formal, so as to give adequate guidance for policies, strategies and plans for environmental sustainable within the country.

On the other hand at this stage includes the identification of the elements that characterize the geological risk and are represented in the bibliographic search and mapping, GIS and geological engineering reports. Is introduced as a factor in the social use of the work, for logical reasons to the determination of influence of the same on the geological environment, dynamic and static loads on the ground, pollution load, etc. Note that as a tool mention GIS also can be used as previous research document or possess the scope to address the task, irrespective of those made specifically for investment in the implementation plan. It is significant to note that reading about the vulnerability and the risk of geologists, geophysicists, hydrologists, engineers, planners, etc. can be very different from reading with people and communities at risk. It is therefore necessary to deepen also the knowledge about individual and collective perception of risk and to investigate the cultural and organizational development of companies that promote or impede the prevention and mitigation; aspects of fundamental importance to find efficient and effective means to succeed in reducing the impact of disasters caused by geological events.

Throughout the construction process, the rules have some point of application, it is necessary from this stage to identify what those involved in knowledge management and apply them properly, a key objective diagnosis. These are issues that should appear reflected in the records of the process.

DESIGN PROCESS

The objective of this process is to establish the rationale and technique to be developed on the various projects of knowledge in the organization. Includes the following:

- **Developing a Knowledge Strategy:** Aimed at setting the course to enable the organization to go from current state to desired state. Aims to establish development plans and project management.

- **Defining a Strategic Goal**: It aims to set the address to which projects are targeted. For a goal is met, must have the following characteristics:

Once defined, the goal should be broken down into objectives, depending on the level of performance to be raised. If the goals are verifiable, they should explicitly presents the achievements and deadlines to be met, i.e. should be described in terms that will generate strong indicators for assessing the associated implementations. Also bear in mind the context that explicitly defines the vision, goals, and corporate philosophy that represents the entire organization.

Corresponding to this is accomplished by designing architecture of knowledge: in order to establish elements:

- **Investments in Technology**: identifying the needs-oriented support model components.
- The patterns of development or integration of the management

model of geological risks: establish guidelines for the development and integration of knowledge management to support the process of geological risk management.

- **The Architecture of the Model Diagrams**: organization and structure of quality control systems to support the model components.

- **The Organizational Climate**: aims to support strategically by management: the expected benefits, objectives and assumptions, developed strategy and its measures, and achieved expected results.

- **Training**: preparing scientific and technical staff who will speak both in execution and assessment processes and control provided for in the model.

At this stage, proposed to the specific management of geological risk carrying out the following:

- Preparation and delivery of Geological Engineering Task. Study of physical-mechanical properties of soils and its relation to the information obtained earlier, the behavior of the project and surrounding loads, analysis of the geology and geo-environmental situation in general. This includes the analysis of geological engineering report updated taking into account variations in the behavior of soils and rocks, topography and other factors changing over time. (geotechnical risk management).

- Identification and review of the main rules governing the implementation of these activities.

- Preparation of preliminary report concluding geological risk management, which must include the results of all investigations in the field of engineering geology, performed either by design engineers, for companies providing geotechnical services as well as their interpretation in terms of the work to be executed. This report provides a basis for making decisions necessary for the design and early implementation of the project, which include mitigation measures preliminary geological risk.

- Assessment of cognitive development achieved by the technical staff on the geological risk, through different techniques.

- Preliminary assessment of the effectiveness of the comprehensive measures taken on the basis of the work designed, geological and social environment.

- Fabrication of the chips in the process, explained the contents and tasks of each thread of the model for this stage, aimed at reducing vulnerability constructive.

It is clear that the availability of information resources in a project does not necessarily guarantee the perfection of its use. The biggest problems are directly related to the effectiveness and efficiency of use and information management is the absence of their organization or their inconsistency. In line with this reasoning, we must consider the possible establishment of an internal program within the implementing institutions or companies to elevate the culture of information, questions relating to the necessary ongoing training of professionals and specialists.

It is possible that information obtained in the literature search, obtained the necessary elements enabling the designer himself prepare a geotechnical report for the work. It happens that in the archives of the institutions are the reports of previous works performed in the study areas, however, are not used, combine economic and administrative procedures unnecessary expensive single project.

According to Ayala (1992) to establish the types of geological hazards in any area in question is first necessary to establish the geological setting, which is basically a study morphological, sediment logical and tectonic elements focused on morph and establish a territory morph structural and its genesis, to establish the stratigraphy and sedimentary faces sedimentary tectonic structures identify, type, address and occupation thereof, and to know the physical properties of the geotechnical and geochemical types of sediment (soil). This geological setting allows identification of potential geological hazards.

Nevertheless, other factors also are factors to assess, these are the geo-environmental factors. The determination of geo-environmental factors, such as the presence of sedimentary instability, erosion and sedimentation rates, bottom currents, fluid dynamics, influence of atmospheric phenomena on the change of geological conditions, presence of gas, gas hydrates, etc.., helps to make a risk analysis with great precision.

The integration of the results obtained in the field of geological setting and geo-environmental factors can assess risk, in terms of frequency, extent affected by the risk and possible pollution due to the outbreak of the geological risk. This line of approach is relating to prevention efforts and agrees with the approach established by the Disaster Mitigation Program of the Agency for the Coordination of United Nations Disaster Relief-UNDRO[6,] and the scientific community.

As part of the successful completion of the model at this stage, ideally, for example, to flood areas with a certain lithology, there was an internal regulation of provincial construction group, to guide the builder types of foundations to be used, height must presented the beginning of the useful structure of the building or work of infrastructure, among other parameters. Process to be carried out by integrating all types of geological hazards present, so as to ensure effective mitigation of risk. These are matters which have already been working in the country by other specialists, and therefore are not analyzed in this research, even if their validity as example of concrete action.

In case of special requirements in the work to execute, it must request special reports geological risk assessment to institutions that specialize in such services, and implement the required mitigation measures and the preliminary evaluation of its effectiveness.

DURING THE EXECUTION (PART INITIAL IMPLEMENTATION)

Implementation Process

This process aims to implement the project and establish its basic guidelines.

Includes:

- **Implementation Plans Developed**: Each of the projects must be implemented according to schedule or plan.
- **Strategy Review**: should be reviewed periodically, both goals and the objectives and plans associated with the strategy.
- Fabrication of the chips in the process, explained the contents and tasks of each thread of the model.

Depending on the geological risk management is proposed to undertake the following actions:

- Practical implementation of the mitigation measures planned for the project geology in the preliminary report prepared in the previous stage.
- Realization of monitoring compliance with the technical and scheduled tasks on technological and productive processes designed in the project.
- Continuous evaluation of a system of indicators for project implementation, to ensure the management of geological risk at this stage of the work.
- Preliminary assessment of the effectiveness of risk mitigation measures taken on the basis of geological work in the implementation process.

At this stage, which falls during the execution of the project, geological risk management is closely related to geotechnical testing, i.e. the physical-mechanical properties of the soils and rocks under study and the implementation of mitigation measures determined in preliminary studies in the first stage.

An indicator allows monitoring and periodic evaluation of key variables or indicators of risk management through comparisons with their internal and external referents. The indicators are evaluated must also see to the implementation of environmental or ecological traps, and techniques during the process of land preparation, proper authorization and rehinchos stuffed, cut and natural slopes, design and implementation of the excavations, the design and implementation of foundations, hydraulics, electrical, and others who will be buried, to carry out works of protection, quality completion of phases, including:

- Human alterations of the landscape.
- Induced instabilities and landslides.
- Changes of content in phreatic level and humidity.
- Observation onsite of the behavior of charges projected onto the soil.
- Changes or variations in the initial design of the project.
- Analysis of soil conditions in buildings in different processes. (Referring to works that are the subject of rehabilitation, remodeling or maintenance, changes of uses of works or objects of work, and changes in environmental conditions, etc.).

There is no set limit on the evaluation of indicators, the more they are the better for the work, as they may be risk factors if not taken into account during the execution of it, a more complete description of these aspects is not objective of this research, this is a task that construction specialists have ahead to solve.

PROJECT COMPLETION STAGE (FINAL PART OF IMPLEMENTATION)

At this stage not "last", it stops being less important the task of managing geohazards in it there are certain actions that also need to look closely as practical experience and visual observation made for this research, need to say so:

- Assess the implementation of a system of indicators for the completion of the project investor, to ensure the management of geological risk at this stage of the work.
- Final evaluation of the effectiveness of risk mitigation measures taken on the basis of geological work in the process of completion.
- Prepare recommendations for efficient operation of the equipment installed in compliance with the geological aspects.

At this stage of completion, the assessment is related indicators, for example, the analysis of the atmosphere works. Many times in our buildings do not work correctly applied the appropriate atmosphere and garden, and we do not foresee the future risk that they may bring works in progress of completion, partly for lack of knowledge about the physical-mechanical properties of soils insitu and filling by specialists in gardening or background, and partly because of lack of guidance from engineers involved in the execution of the work directly on these issues.

As a result, going from a few years begin in the buildings directly affected by the growth of root systems, which manifest themselves in various ways such as: cracking, subsidence, building of walls, exterior walls and interior structures, and even the collapse of these structures.

These effects occur due to the characteristics of the soil, providing nutrients needed for plant development, a process that often increases as a geologic event occur such as floods or earthquakes, which in the case of first increases in soil properties such as porosity and pore rate, reduces compaction, and promotes the increase of plastic properties and the second, and increase the plastic characteristics of the soil, increases moisture and liquid content, being all these consequences, favorable factors for the development of plants and their root systems (root length, nutrient solution and water for its development.)

To avoid these phenomena is also necessary to manage this risk in the process of completing the work, and even promote actions for the cognitive growth of these elements in setting workers, and technicians and engineers about their work, so that there is a feedback between them.

It is necessary for having gained extensive knowledge in the course of the previous stages, and is even on the geological risk and how they managed correctly for each construction, infrastructure work or subject-specific work, this is described by a report and given to future utility, with a set of recommendations for the efficient operation of the equipment installed in compliance with the geological aspects are identified and managed during execution.

This is a very important for construction companies, allowing them to among other things, take a highly technical and responsible position in the field of geological hazards. On the other hand, it also permits the protection of technological knowledge for the perpetrators, protect its reputation in the community and contribute to successful and appropriate future use of the work in question.

CONTINUOUS IMPROVEMENT PROCESS (PROCESS EVALUATION)

Evaluation Process

Its objective is to assess the results of the implementation of projects, to validate the strategy of knowledge and diagnostic feedback to the process. This process provides that, once the implementation of projects and their plans, they must be evaluated by a number of management measures, and this will show the results in the incorporation of mitigation measures in the context the project.

It should be narrow, that each completed project is a virtual laboratory for the construction company, as the continuous improvement includes both aspects of this work as benchmarking with other works carried out. Allow internal comparisons show the progress from the historical perspective of the vision of designer and executor. However, a comparison with the outside will show the real impact of progress, because it allows comparison of the relative effectiveness in the management of geological hazards.

To perform these evaluations, different modalities can be applied:

- **Quantitative Measurements:** pre-defined variables and that have meaning.
- **Qualitative Measures**: through non-numerical methods.
- **Observation:** corresponds to the views of the staff previously trained to evaluate issues of concern.

This process includes:

- **Implementation of the Measurements**: Definition of method and technique to obtain information and execute measurements according to the defined actions to obtain the necessary information.
- **Interpretation of Results**: Includes the processing and analysis of the data to determine the type of geological hazard for which the indicator was created. Depending on the volume of information can be validated using the selected tool.
- **Continuous Improvement**: involves applying the principles of analysis provided for in Denim cycle, consisting of evaluating potential errors or improvements to the procedures, techniques or technologies used in the work. This technique applies both during implementation and during operation, and it aims to improve all processes running on the play in terms of both itself and improve them in future projects.

Under this conception, the works carried out are the practical laboratory of companies implementing or controlling the processes of geological risk management. This still enables the conduct of mitigation measures applicable to future projects, and improvements in the works already completed or in process maintenance, rehabilitation or remodeling.

GRAPHICAL EXPRESSION OF THE GEOLOGICAL RISK MANAGEMENT MODEL FOR CONSTRUCTION AND INFRASTRUCTURE PROCESSES

After having described the steps that will be present in the model, having made a thorough analysis of the elements that make up the geological risk in particular, to analyze further how day geological risk is managed from the point of view institutions and legal regulations involved in this process and describe the procedures of the model, it is appropriate to make a graphical representation of it:

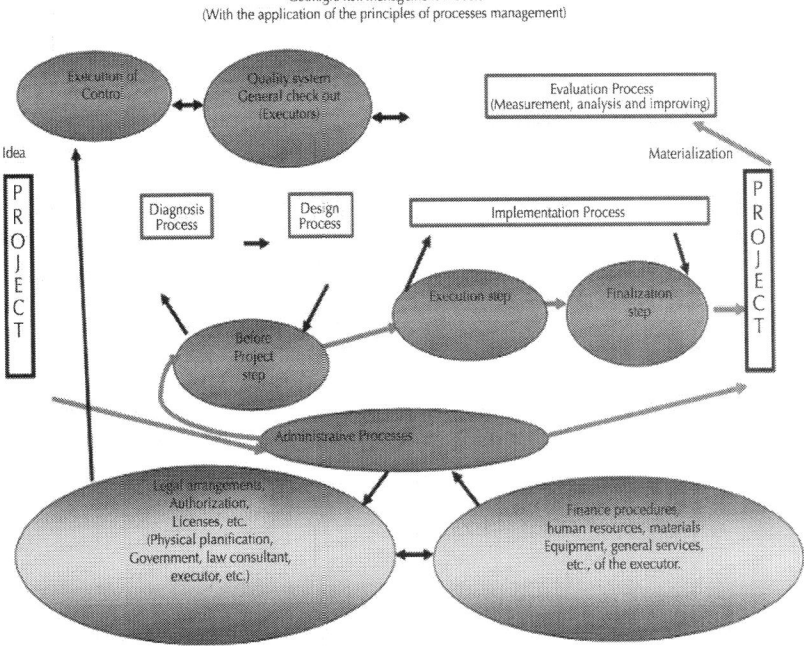

Figure 6: Model for geological risk management in construction and infrastructure processes.

The model explains three types of fundamental processes:

- The administrative, that are intended to ensure the management of human resources, financial and material (bottom).
- The principal managers of the geological risk: which are interconnected by the methodological steps proposed in this research.
- The executive control and continuous improvement: for the control that those responsible companies and engineers are executing on the execution of all activities related to the project, geological risk management on an ongoing basis (at the top).

We believe it is significant to state that this model is a dynamic model. The processes indicated in the management, may or may not be applied in correspondence with the type and size of investment, as well as the assessment is made of time of use of the work. Just as other aspects may include consideration of the executor are important risk factors to be managed.

It can see that the process of geological risk management is a complex process where several factors, which in the end always help to assess the magnitude of risk and vulnerability and cumulative, and in this way make timely technological measures necessary for success, durability and security of investment and infrastructure construction, which is printed therefore more sustainable development constructive and life safety.

The successful implementation of the model lies not only in knowledge of the steps, stages or threads that shape it, but also in the interpretation efficiency geodynamic situation described and provided by the various documents, information contained in GIS, tabs and others, who are able to obtain the executing engineers and investors in the region, area or locality where the work is located and also the correct application of the rules, regulations and technical measures contained in the various ministries, regardless of its shortcomings.

Each of the above analysis involves uncertainties and limitations that are reflected in the final application of mitigation of geological risks, which must be taken into account when interpreting the results of the staff responsible for implementation. These potential limitations include:

- Inadequate search of information needed to manage the geological risk.
- Do not apply the provisions of the various building regulations, rules geotechnical, seismic or other.
- Do not apply the issues raised in the various resolutions, plans and regulations currently existing in the environment, technology, civil defense, construction, etc.
- No other measures that are not listed in these sections and which may form part of the introduction of new technologies either by transfer or by innovations made in connection with the work.
- Do not make the necessary executive and technical control compliance activities described in the management of geological risk for processes and threads.
- Do not apply the different variables are and can be introduced to the model in correspondence with the type, nature and extent of the constructive or infrastructure to run.
- Failure to make a good staff training on issues related to both the geological risk management, and the application of the model during the construction process.
- Do not consider the process of continuous improvement as part of the management of geological risk.
- Not adequately prepare the files of the processes and under process.

The limitations to the application of this model are highly dependent on subjective factors that have to do with the knowledge to meet the task ahead, preparation of staff and with effective control of the actors of the investment process. This is the key to ultimate success and quality assurance in the implementation of geologic risk mitigation measures.

METHODOLOGY OF IMPLEMENTATION OF THE MODEL

To implement the management model of geological risks in an organization requires the implementation of a methodology and strategy. Implementing the strategy involves conducting a series of actions will be met through the methodology and procedures established for that purpose. This strategy is tailored to each organization in correspondence with the analysis of the factual situation that is real, which means that its implementation depends on the internal characteristics of the organization, its corporate purpose, level of training of staff to plunge homework, etc.. The strategy should also optimize the balance between quality, time and cost, according to the priorities assigned to each of these variables.

The implementation methodology consists of four steps, differentiated by the objective pursued in each of them:

- **First**: Identifying and assessing the current state: it corresponds with the diagnosis and inventories of resources and services that are available both as identifying those that can be implemented through the implementation of various projects.

- **Second:** Definition of goals: establishing a diagnosis made according to and knowledge of organizational behavior. Therefore, as part of the design process, it is proposed to implement the model for the management of geological risks so as to focus its efforts in the allocation of content that realistically reflects the potential of knowledge within the organization.

- **Third:** Project development: it takes place after the implementation of the actions of the strategy designed for that purpose and which will gradually incorporate mitigation measures geological risk, as structured in different phases and applications to express knowledge of the organization and its relationship with the environment.

- **Fourth:** Analysis of results: examines the correspondence between the results of determining the current state with the goals that define the organization and the definition of the projects carried out to establish the differences that must be given a new diagnosis.

As an indispensable element and prior to the successful implementation of the methodology, it should ensure the effective engagement of the direction of the company as a rector of any change, and employees as direct and decisive factor in realizing the process improvement.

Set to the methodology steps 1, 2 and 3 are made by members of the management of the company, which will oversee the overall development of it.

The first step of the methodology should provide inter alia for internal and external analysis, which comes from the direction of the company, and where research should focus broadly on what factors are influencing the actions of the system, identifying results, effects of daily management, etc. This will differentiate the results that are the product of external factors and those from internal.

For this analysis should be selected and formed an interdisciplinary team which has the following characteristics: (Negrin, 2006)

- Consist of between seven and 15 people. ((Recommended 9)
- Ensuring the diversity of knowledge of team members.
- Some of the members have to be experts in management systems
- Having the presence of an external expert on geological hazards.
- Appoint a member of the Management and Coordinator of the team.

The proposed technique for the analysis is brainstorming, which will be held for each functional area and level of the entire project. This step will be the starting point for an analysis of the processes inside the company and will detail the problems of each process, based on the application of the methodology, and will identify whether the factors that must be improved causal relationship on the effects or results of the geological risk management.

To identify and define the goals in the organization is necessary to take into account certain issues:

- There must be a contrast with the strategic objectives of the company.
- Must meet the needs of clients of the investment process, understanding as such all persons or entities own or outside the company, which receive some of the outputs of the process.
- They must meet the expectations of the management process of geological risks during the construction process and infrastructure.
- Should be addressed to improve in the management of the final quality of the work and the recognition in the community of the administration by the company executing the project.
- Should take into account material weaknesses and problems related to human resources.

During the development of the project the team designated to carry out each process, thus arises the need to define indicators for geological risk management in response to the following questions:

- What should we measure?
- Where should you measure?
- When should I measure? At what time or how often?
- Who should measure?
- How should you measure?
- How do they have to disseminate the results?
- Who and how often you will review and / or audit the data collection system?

Where the first work to be done with these indicators is to realize the objectives of all the indicators defined in the previous phase, so these are consistent with the basic objectives of the process and ensure compliance.

As discussed above, the model suggests some of the key indicators for the management of geological risk, of course the objective reality of the construction process, the company and the environment, will show which of these are the most efficient, and if described in the model are appropriate or necessary to its growth.

The process should be evaluated periodically. This is a very important aspect is often forgotten by staff blamed in developing this type of activity. The assessment of performance of a process, reference must be made on a pattern of functional excellence This pattern of comparison to be made from desirable or optimal behavior of a set of measuring the performance of processes in the world's leading construction companies, or alternatively in the Cuban construction companies with similar processes in order to study, with proven success in performance. All this by a synthetic indicator, which when calculated in quantitative terms to identify the gaps between the actual level of the meters and their desired trend, which makes it possible to define specific problems in all dimensions of the process.

An important issue that can effectively introduce mitigation of geological risks in construction projects and infrastructure construction is the evaluation and selection of alternatives for improvement. Process Team to evaluate possible actions to take to solve the problems that have the greatest impact on the performance of the process, taking into account the feasibility of comprehensive implementation and its impact on the whole system under these conditions prepares a draft improvement plan with responsibilities and deadlines, in order to define and validate how to implement the improvement, that is the measure of geological risk mitigation.

To efficiently solve the latter issue, the process analysis team may use some of the tools provided in the following management processes:

- **Troubleshooting:** This application is applied locally to the selected activities as long as the information is specific enough to describe the object or location is detected and the particular defect that occurs. Any tool related to the resolution of problems is valid.

- **Technical Value**: To identify possible wastage of the current process, we proceed to apply this technique to all the process activities identified with some degree of difficulty, systematically questioning all of them. Be sufficient to make the following questions in a first approximation (If necessary, resort to using the tool in all its depth):

Does it contribute to increase the quality and safety of the work?

Does it contribute to meeting the needs of the client or investor?

Does it contribute to achieve one of the strategic objectives? etc.

Gather external information related to the process or activity thereof. Depending on the extent of the process may be interesting to divide the work of capture and analysis of information between different team members, the sources indicated in the model.

After selecting or selected improvement alternatives, it is necessary to establish the improvement plan at this stage is part of the results of which have been defined above problems have a greater impact on the process individually and as expanded on the strategic objectives of the companies and also the real possibility of giving a viable solution for the company in the short term, so we proceed to define a plan of improvements to the final process with the highest degree of detail, which include action to take, material resources, financial and human resources to employ, directly responsible for implementing the improvement and the impact this will have on the process and organization. To implement the improvement alternatives is implemented the improvement plan previously defined, the implementation may last longer, so it is necessary to develop a concrete plan with defined responsibilities, deadlines for each of the objectives of proposed improvements.

The implementation phase of improvements to the process requires that the Department will approve the proposed direct interaction with all workers involved in the process (workers, technicians and managers). Before implementing the new process is necessary to think about possible resistance change and possible countermeasures to be adopted among which are the following: (Negrin, 2006)

- Communicate and involve people who will be involved in the implementation of improvements.
- Provide education and training necessary
- choose the right Timing
- Develop a progressive implementation of improvements, trying to start this with the most receptive and the most prestigious among their peers.

Prior to implementation, is introduced to the company's usual (procedures, instructions, rules, etc.) Changes associated with improvements in order to consolidate the changes and avoid internal contradictions.

May be used in the analysis of results, different methods of assessment have been described in the literature, such as SWOT matrix to facilitate knowledge management.

The steps of the methodology are not closed systems, but are enriched by the ideas, according to the needs of each organization in which methodology is used, but always considering that the instruments are implemented and planned actions that respond to the objectives sought in each, these established principles of process management.

To monitor and evaluate the results, the improvement team responsible driving the implementation of the Plan of Implementation, compliance controls and evaluates the effectiveness of the work done by monitoring the results achieved, and performing periodic filings with the management of the company, head of compliance plan process improvements.

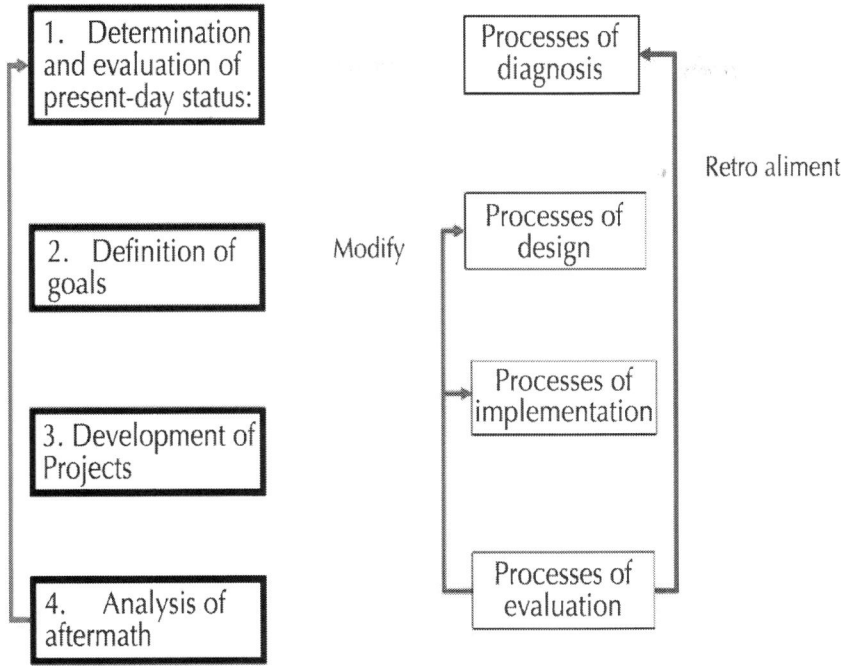

Figure 7: Structure of the methodology.

The application of the methodology is systemic and cyclical nature because the assessment is a diagnostic feedback and modifies the actions to be undertaken, both from the point of view of knowledge as its architecture as shown in Figure 7.

The application of the geological risk management under the principles of process management is also cyclical raw satisfaction of stakeholders in the final product that our case is the constructive or infrastructure quality in the final involving mitigation of geological hazards and the improvement process.

CONCLUSIONS

The model aims to provide new ideas on the management of geological risks in the construction and infrastructure processes, so that as a proposition is introduced into the investment process is carried out in different countries, from its legal body, in a manner that establish a requirement for active and passive actors in this process, this would be in the author's opinion, the best way to establish a way to ensure sustainability of building development in the medium and long term. This also would avoid the large outlay of money and resources that the states have to pay or release each year, after the frequent occurrence of natural and technological disasters.

It is relevant to mention that as a model of technological innovation, this model is dynamic, allowing for interaction between actors and resource persons. Its main limitations depend on how deep or not be covered by the interdisciplinarity participation, to what extent would be apply or not the different management tools, as well as aspects related to the introduction of technology transfer and / or innovations in the process.

The methodology presented to validate the model, start of the criteria expressed by the experts consulted during the research, the actual situation of the companies executing projects in Cuba, which are introducing gradually Principles of process management into their systems and subsystems.

The methodology raises the application of the model with a cyclical and interactive character, where is prevailing the training and group decision, and the individual insistence in administrative control at every stage of the work and, overall, corporate actions, which are developed to finally make a proper geological risk management in the constructive and infrastructure processes.

REFERENCES

1. Ayala 1992 1992. Introducción a los riesgos geológicos. Instituto Geológico y Minero de España. Editorial Ríos Rosa, 23. 28003. Madrid.

2. J. Yolcina, 2002. Riesgos Naturales. Editorial Ariel, Barcelona. 8-43448-034-4

3. Bieri, 2005Disaster Risk Management and the Systems Approach. http://www.drmonline.net./drmlibrary/pdfs/systemsapproach. pdf. Consultado en Diciembre 2007.

4. O. Cardona, A. Darío, 2001 Evaluación de la amenaza, la vulnerabilidad y el riesgo. "Elementos para el Ordenamiento y la Planeación del Desarrollo". Red de estudios sociales en Prevención de desastres en América Latina. Bogotá. Colombia. http://www.desenredando.org/public/libros. Consultado en Junio del 2006.

5. R. Chuy, J. Tomás, Puente, 2005 Impacto de fenómenos naturales. Una valoración imprescindible para el desarrollo sostenible de zonas costeras de Santiago de Cuba. Obtenido en formato electrónico en Centro de Estudios de Manejo Costero. Universidad de Oriente. Santiago de Cuba.

6. Colectivo de autores, 2005 "EXELENCIA EMPRESARIAL, Por qué la gestión por procesos", España. http://web.jet.es/amozarrain/index.html.Consultado en Marzo 2008.

7. Colectivo de autores. 1999 Tecnología y sociedad. Grupo de Estudios Sociales de la Tecnología (GEST). Editorial Félix Varela, La Habana. Cuba.

8. Colectivo de autores. 2005 Guía ambiental para obras de prevención y mitigación de riesgos. Quinta Parte. Organización Panamericana de la Salud. Biblioteca virtual de desarrollo sostenible y salud ambiental. www.bvsde.paho.org/bvsacd/cd65/GuiaAmbiental/biblio.pdf. Consultado en octubre 2007.

9. Concepción 2003 Metodología de Gestión de Proyectos en las Administraciones Públicas según ISO 10.006 Localización.http://dialnet.unirioja.es/servlet/oaites?codigo=1434. Consultado en octubre 2007.

10. V. Crespo, D. Carlos, 2004 Mecánica de suelos y cimentaciones. 5ta.Edición. Editorial Limusa, Noriega editores. España. 9-68186-489-1 Pérez-Villar. El concepto de Gestión del Conocimiento. http://www.gestiopolis.com/canales6/ger/la-gestion-del-conocimiento.htm, Consultado en noviembre 2007

11. E. M. Fournier, 1985 "The Quantification of Seismic Hazard for the Purposes of Risk Assessment", International Conference on Reconstruction, Restauration and Urban Planning of Towns and Regions in Seismic Prone Areas, Skopje. Obtenido en formato electrónico.

12. Administración. Fundamentos de, J. Financiera, Weston. Fred, F. Eugene, Graw. Brighan-Hill Mc, 1994 Galbán Rodríguez,

Liber(1). 2009. Algunas consideraciones sobre la introducción de las nuevas tendencias internacionales en materia de gestión de riesgos geológicos, en la enseñanza de la ingeniería hidráulica y ambiental en Cuba. Ponencia presentada el 7mo. Congreso Provincial de Educación Superior.

13. Universidad 2010 Junio, 2009. Palacio de Convenciones Heredia. Santiago de Cuba. Cuba.

14. Galbán 2009 Algunas consideraciones teóricas sobre la gestión de riesgos geológicos. Revista de Geología UFC. Volúmen 22, Número 1. Brasil. 0103-241 0103 2410

15. Galbán (3). 2009. Algunas reflexiones sobre las causas que generan el riesgo geológico en la provincia Santiago de Cuba. CD ROM "III Taller Internacional Nuestro Caribe en el Nuevo Milenio". 978-9-59207-357-9

16. Galbán (4). 2009. El modelo de gestión por procesos en la evaluación de riesgo geológico en la provincia Santiago de Cuba. Un ensayo preliminar. Rev. Mapping, 1131-9100 1131 9100 , N 132, 2009, pags. 18-23. España. http://dialnet.unirioja.es/servlet/articulo?codigo=2913108 Referenciada en: LATINDEX, COMPLUDOC, DIALNET.

17. Galbán (5). 2009. Modelo para la gestión del riesgo geológico en los procesos constructivos y de infraestructura. Revista de Obras Públicas: Organo profesional de los ingenieros de caminos, canales y puertos, 0034-8619 0034 8619 , N. 3500, 2009, pags. 39-50. España. Referenciada en: COMPENDEX, COMPLUDOC, GEOREF, ISOC, ICYT, LATINDEX, TRANSPORT, TECNOCIENCIA, DIALNET.

18. Galbán (6). 2009. Problemas sociales que enfrenta la gestión de riesgos geológicos en los procesos constructivos y de infraestructura en Cuba.http://www.monografias.com/trabajos75/problemas-sociales-gestion-riesgos-geologicos/problemas-sociales-gestion-riesgos-geologicos.shtml

19. Galbán (7). Conferencias del Curso Geología para Ingenieros. 2008. http://webserver.fco.uo.edu.cu/uoclas/LGR

20. R. Galbán, Galbán R, Liuba; P., Ársul J.; Gago A. 2010. Reflexiones en materia de gestión de riesgos geológicos en procesos constructivos del municipio Santiago de Cuba: Normas

y procedimientos jurídicos. Revista Jurídicas. 27 3, 2010. Colombia. 1794-2918

21. V. Iturralde, A. Manuel, R. González, E. Bertha, R. Chuy, R. Tomás, naturales de origen geológico. 2006www.medioambiente.cu/ uptnatgeo/index1.htm. Consultado en diciembre 2007.

22. E. Keller, 1995 Enviromental Geology. Prentice may, New Jersey, 560pp.

23. Kiroiwa 2002 Reducción de desastres: Viviendo en armonía con la naturaleza. Editorial Quebecor World, Perú S.A. 997294770

24. H. Koontz, H. Weihrich, Administración una prospectiva global Editorial-Hill Mc Graw edición, México, 1998.

25. M. Y. Dikdan, S. Jaua, 2003 Modelo de aseguramiento de la calidad en el diseño y construcción de desarrollos masivos de viviendas de interés social. VII Congreso Latinoamericano de Patología de la Construcción y IX Congreso de Control de Calidad en la Construcción. CONPAT 2003, Vol. I: Control de Calidad, Capítulo VIII : Gestión, Trabajo VE01 pp. VIII. 9 - VIII. 16. 9-68464-133-8 Yucatán, México

26. Jauge. D. Mendioroz, 2003 "Gestión Integral de Obra",III Congreso Andaluz de Carreteras. España. Obtenido en formato electrónico en Centro Territorial de Gestión de la Información del MICONS, Santiago de Cuba.

27. Granados. Monge, 2003.". Hernando, 2003."La construcción de proyectos de infraestructura multinacionales en Centroamérica y sus consecuencias en la generación de riesgos". Costa Rica. Obtenido en formato electrónico en Centro Territorial de Gestión de la Información del MICONS, Santiago de Cuba.

28. National Academy Of Sciences, Earthquake Prediction and Public Policy, Commission on Sociotechnical Systems, National Research Council, Washington, 1975.

29. Ernesto. Negrin, 2006 "Metodología para el perfeccionamiento de los procesos en empresas hoteleras". Consultado en noviembre 2008.http://www.monografias.com/trabajos10/hotel/hotel.shtml.

30. Norma ISO 90012001 . Obtenido en formato electrónico en Centro Territorial de Gestión de la Información del MICONS, Santiago de Cuba, 2008

31. J. Perkings, , «Liability of Local Government for Earthquake Hazards and Losses- A Guide to the Law and its Impacts in the States of California, Alaska, Utah and Washington», ABAG, Oakland, 1989.

32. L Quarantelli, 1992. Urban vulnerability and technological hazards in developing countries societies. Washington DC. USA. Obtenido en formato electrónico en Centro Territorial de Gestión de la Información del MICONS, Santiago de Cuba, 2008.

33. R de Calvío Fundasal, , Z Gilma, . 2005. Hacia una metodología para la gestión del riesgo en comunidades marginales. Consultado en octubre 2007. www.yorku.ca/ishd/RICOdeCALVIO.pdf.

34. MA Soto Balbón, , NM Barrios Fernández, . 2006. Gestión del conocimiento. Parte II. Modelo de gestión por procesos. Acimed.http://bvs.sld.cu/revistas/aci/vol14_3_06/aci05306.htm. Consultado en Julio 2008.

35. R.J.S Spence, 1990.»Seismic Risk Modelling - A review of Methods», contribution to «Velso il New Planning», University of Naples, Papers of Martin Centre for Architectural and Urban Studies, Cambridge. Obtenido en formato electrónico en Centro Territorial de Gestión de la Información del MICONS, Santiago de Cuba.

36. C. Starr, 1969 «Social Benefit vs. Technical Risk», Science, American Association for the Advancement of Science, Vol. 165, Sept..

37. UNDRO, «Natural Disasters and Vulnerability Analysis», Report of Experts Group Meeting, Geneva, July 1979.

38. A. Zucchetti, et al. 2008. Guía Metodológica para el Ordenamiento Territorial y la Gestión de Riesgos. Equipo Técnico Grupo GEA. www.grupogea.org.pe. Depósito Legal: 2008-05506, HS Number: HS/983/08S, ISBN Number:(Volume) 978-92-1-131966-8. Lima, Peru.

chapter 4

Adaptive Governance and Institutional Strategies for Climate-Induced Community Relocations in Alaska

Robin Bronen[a,1] and F. Stuart Chapin[b]

[a]Resilience and Adaptation Program, Alaska Institute for Justice, and
[b]Institute of Arctic Biology, University of Alaska, Fairbanks, AK, 99775
[1]To whom correspondence should be addressed.

ABSTRACT

This article presents governance and institutional strategies for climate-induced community relocations. In Alaska, repeated extreme weather events coupled with climate change-induced coastal erosion impact the habitability of entire communities. Community residents and government agencies concur that relocation is the only adaptation strategy that can protect lives and infrastructure. Community relocation stretches the financial and institutional capacity of existing governance institutions. Based on a comparative analysis of three Alaskan communities, Kivalina, Newtok, and Shishmaref, which have chosen to relocate, we examine the institutional constraints to relocation in the United States. We identify policy changes and components of a toolkit that can facilitate community-based adaptation when environmental events threaten people's lives and protection in place is not possible. Policy changes include amendment of the Stafford Act to include gradual geophysical processes, such as erosion, in the statutory definition of disaster and the creation of an adaptive governance framework to allow communities a continuum of responses from protection in place to community relocation. Key components of the toolkit are local leadership and integration of social and ecological well-being into adaptation planning.

INTRODUCTION

Human displacement could be a severe humanitarian consequence of climate change (1). Natural disasters have increased substantially over the past century, with ~370 natural disasters (more than one per day) displacing 38 million people in 2010 (2, 3). Floods caused 182 of these disasters, affecting 180 million people and killing 8,100 (2).

Approximately 10% of the world's population resides in coastal communities that are 10 m or less above current sea level (4, 5). The complex interplay of repeated extreme weather events and on-going biophysical processes, such as erosion and climate-induced sea-level rise, may permanently displace the inhabitants of many coastal communities, particularly in low-lying island nations (6), subsiding river deltas (7), and zones of active coastal erosion (8, 9).

Disaster relief and hazard mitigation are the traditional humanitarian responses to extreme environmental events and are primarily aimed at rebuilding and repairing infrastructure in place and protecting them from future hazards (10). However, this approach may be futile when climate change-induced biophysical changes repeatedly alter ecosystems, damage or destroy public infrastructure, and endanger human lives (11), in which case community relocation involving permanent population displacement may be the only viable adaptation. Climigration is a specific type of permanent population displacement that occurs when community relocation is required to protect residents from climate-induced biophysical changes that alter ecosystems, damage or destroy public infrastructure, and repeatedly endanger human lives (11). In this context, community relocation includes the reconstruction of livelihoods as well as the rebuilding of housing and public infrastructure in a location, away from vulnerable risk-prone coastal and riverine areas. Such relocation provides an opportunity for planned retreat from untenable situations.

In the United States there is currently no institutional framework or agency with the authority to relocate the entire public and private infrastructure of a community and rebuild livelihoods in a new location to protect them from climate change-induced hazards (10). Determining appropriate adaptive responses requires a sophisticated on-going assessment of a community's social, political, and economic susceptibility to harm caused by climate change and its capacity to adapt through protection in place, managed retreat of some structures, or community-wide relocation. There is currently no legislation authorizing funding for such assessments.

Climate change already impacts the habitability of many Alaskan communities. The US Government Accountability Office found that flooding and erosion affect 184 of 213 of Alaska Native villages (12), with 31 of these imminently threatened, and 12 communities planning to relocate (10). [Throughout this article the term "village" refers to an Alaska Native community: (i) deemed eligible as a Native village under the Alaska Native Claims Settlement Act; and (ii) which has a corresponding Alaska Native entity that is recognized and eligible to receive services from the Department of the Interior's Bureau of Indian Affairs (10). The term "community" is used more broadly to describe Alaska Native villages as well as other population aggregations defined by geographic proximity.] Despite state and federal expenditure of millions of dollars, erosion control and flood protection have not been able to protect some communities. The inability of technology to protect people who reside in vulnerable risk-prone coastal and riverine communities could affect millions of people globally. The 2012 devastation caused by Hurricane Sandy exemplifies these risks. The state governments of New York and New Jersey are now evaluating whether rebuilding coastal communities is possible and whether erosion and flood control infrastructure can protect these communities in the future (13).

This article describes the Alaskan experience with these issues. For several Alaska Native communities protection in place is not possible, and communities and government agencies agree that relocation is the only adaptation strategy that can protect them from accelerating climate-change impacts. We first discuss the suitability of the current postdisaster and hazard-mitigation statutory framework to address climigration in the United States. We then examine the institutional challenges faced by Alaskan communities seeking to relocate in response to climate change. We conclude by describing an adaptive-governance strategy that can provide a continuum of responses from protection in place to community relocation and would allow more effective and less costly adaptation to climate change. Finally, we suggest some policy changes to implement this strategy.

RESULTS

Policy Analysis: Postdisaster and Hazard Mitigation Statutory Framework

Significant statutory limitations prevent the government from responding effectively to the gradual biophysical changes that force communities to relocate in Alaska. The Federal Emergency Management Agency (FEMA), whose activities are defined by the 1988 Stafford Disaster Relief and Emergency Assistance Act, is the federal agency responsible for hazard mitigation and disaster relief in the United States (10, 14). The act requires a presidential disaster declaration to access federal funding for postdisaster recovery, as well as most hazard-mitigation activities (14). Under the Stafford Act, the President is authorized to declare a disaster for natural catastrophes, such as hurricanes and tornados. Drought is the only gradual biophysical process listed in the statute as a potential catalyst for a presidential disaster declaration (14). Erosion, which is one of the significant hazards faced by Alaskan coastal communities, is not included in the list of major disasters in the Stafford Act (14). Federal resources for postdisaster recovery are primarily intended to help rebuild individual homes in their current location (10, 14).

The Disaster Mitigation Act of 2000 modified the Stafford Act by establishing a federal program for predisaster mitigation. Five FEMA grant programs comprise the predisaster-mitigation federal response, none of which provide for community-wide relocation (10). One of the federal hazard-mitigation grant programs, the Hazard Mitigation Grant Program, provides funds to develop a Hazard Mitigation Plan for areas that have been declared a federal disaster (10). Mitigation planning requires a comprehensive risk assessment that helps a community identify and prioritize mitigation activities to prevent or reduce losses from identified hazards (15). Although the regulations require that approved mitigation plans be reviewed at least every 5 y, the integration of this information into risk analyses to inform mitigation activities is costly (15). Funding for mitigation activities is allocated nationally on a competitive basis based on cost-benefit ratios (10). Voluntary property acquisition is one of the tools of the Hazard Mitigation Grant Program to permanently remove structures from floodplains after a disaster has occurred. Homes are individually purchased and demolished or relocated to another location outside the floodplain (16). FEMA recommends that communities not develop relocation sites to which community members can move because of the complexity and expense of the process (16). The program requires that the land in the floodplain be designated as open space for recreational or agricultural purposes in perpetuity after the structures are removed [44 CFR 206.434(d)].

Alaskan communities have difficulty competing for hazard mitigation funds, including the property acquisition program, because of their remote location and low population, which equates to high costs and low benefits (10). In addition, erosion is the primary cause for relocation, and erosion is not included in the list of environmental events, as defined by law, that can initiate a presidential disaster declaration (17). Disaster-relief and hazard-mitigation measures are important when protection in place is possible, but are insufficient to respond to the climate-induced biophysical changes occurring in Alaskan communities.

To respond to this gap, the Alaska State Legislature created the Alaska Climate Change Impact Mitigation Program (ACCIMP) in 2009 to supplement the federal Hazard Mitigation Grant Program (3 AAC 195.040). The ACCIMP provides funds for hazard impact assessments to evaluate climate change-related impacts, including gradual biophysical change, such as erosion. The remaining funds are allocated for the planning needs and adaptation strategies to reduce vulnerability to the hazards identified in these assessments. Relocation planning activities can be funded.

Funding from the ACCIMP is limited to two community categories. Noncompetitive funding is allocated to six communities designated by name that are currently threatened by climate-induced biophysical change. The remaining funds are administered through a competitive grant process to communities based on an evaluation of four factors: (*i*) risk to life or safety during storm or flood events; (*ii*) loss of critical infrastructure; (*iii*) threats to public health; and (*iv*) loss of 10% or more of residential dwellings. The ACCIMP is a government-bridging program that provides a mechanism for communities to assess climate risks and create adaptation strategies, including relocation. However, this regulation does not mandate or authorize any state agency to provide relocation technical assistance, even if relocation is determined to be the most feasible adaptation option to protect lives and property. As a consequence, although ACCIMP allows relocation planning, no institutional relocation governance framework exists to implement community relocation in Alaska.

Community Relocation Efforts in Alaska

Community relocation in Alaska is already a recognized need. In the past, arctic sea ice protected indigenous coastal communities along the Bering and Chukchi Sea from coastal erosion and flooding by creating a barrier to storm-related waves and surges. Regional warming has thawed coastal permafrost because of warmer air and water temperatures (9, 18) and has reduced summer sea ice cover by 39–43% since 1979 (19), leading to a longer fetch and taller waves (20). Together, these changes have increased rates of coastal erosion, especially during severe autumn storms, which (because of the longer ice-free season) are now more likely to occur during ice-free conditions (8,9).

In this section, we describe the relocation process of the three communities identified in the 2003 US Government Accountability Office report as most critical to relocate. The governments of Kivalina, Shishmaref, and Newtok concluded decades ago that community relocation was the only solution to protect their respective communities from life-threatening biophysical change. Each community has undertaken a three-pronged relocation process that involved: (*i*) identification of a new village site, (*ii*) resident voter approval of the relocation site, and (*iii*) documentation to substantiate the need to relocate and the suitability of the relocation site for the community (21–24). Each community commissioned several social-ecological assessments and relocation evaluations. Despite the similarity of the steps taken by each community to relocate, only Newtok has begun the relocation process. A comparison of the three case studies demonstrates a common suite of challenges faced by Alaskan communities seeking to relocate and some of the factors that have either contributed to or constrained progress toward relocation.

The ancestors of the current residents of Kivalina, Shishmaref, and Newtok moved seasonally among coastal and inland hunting and fishing camps (24–27). This migratory lifestyle changed during the late 19th and early 20th centuries primarily because the US Department of the Interior's Bureau of Education began to develop a formal educational system for the Alaska Native community (25,28). The construction of schools along the western coast of Alaska and the requirement that Alaska Native children attend school caused the Alaska Native population to consolidate and settle (25,28). Barge accessibility to transport construction materials determined the location of the schools (25, 26). The building of permanent schools and housing and of sewage, water, and electricity infrastructure led to a change from seasonal migration to establishment of permanent communities at the school sites selected by the federal government (24). This change reduced the flexibility of each community and created a new set of dependencies on government to respond effectively to environmental changes.

Kivalina

The Village of Kivalina is an Inupiaq Eskimo federally recognized indigenous tribe located on the tip of a thin, 6-mile-long barrier reef island in the Chukchi Sea, 128 km above the Arctic Circle (22). Storm surges and flooding threaten the community as a result of diminished arctic sea ice and the delay in freezing of the ocean. Between 2002 and 2007, six extreme weather events threatened Kivalina. The state and federal government issued three disaster declarations (23). The most recent extreme event was a storm hurricane-strength in November 2011 (29). Between 2006 and 2009, government agencies spent $15.5 million on erosion-control projects that have failed to protect the community (23, 30).

Erosion caused by storm surges impacts infrastructure that is essential for the viability of the community in its current location until such time as relocation can occur. These infrastructures include the only means of access to the community (the summer barge landing and the community airstrip), the community's sole water source, and the stability of the community's solid waste storage containment area (22, 23, 27).

In 1998 and 2000, the community voted to relocate and chose two different relocation sites, which the US Army Corps of Engineers (USACE) later determined after each vote were unsuitable because of thawing permafrost (23). In January 2012, Kivalina residents voted to construct a new school 7 miles from their current location. Funding for the new school comes from a lawsuit settlement agreement involving funding inequities that harmed rural Alaskan schools (31). Kivalina's efforts to raise additional relocation funds from a climate-change lawsuit against oil, coal, and gas companies have been unsuccessful. The Kivalina Evacuation and School Site Access Road Committee is coordinating the work to determine the viability of constructing a road between the current community location and the school site. The road will provide an evacuation route during extreme weather, and the school may serve as pioneer infrastructure for community relocation. Funding for the road construction may come from the Alaska Department of Transportation and Public Facilities (DOT) and USACE, but the timing of road construction is unclear. The additional steps required to relocate all of Kivalina's residents, infrastructure, and housing to this location have also not been identified.

Shishmaref

Shishmaref is an Inupiat Eskimo village on Sarichef Island on the northwest coast of Alaska. Between 1973 and 2009, state, federal, and tribal governments invested about $16 million in shoreline protection to address the accelerating rates of erosion (32–34). Despite this investment, storms repeatedly damaged or destroyed public infrastructure and many homes. In 2001, the Native Village of Shishmaref created the Shishmaref Erosion and Relocation Coalition to work with multiple federal agencies and their contractors to identify a new, safe, and culturally appropriate community location (32, 33).

In 2002, residents voted to relocate the community, and two federal government agencies began studying the relocation issue—the USACE, mandated to provide engineering services to reduce risks from disasters, including flood control, and the US Department of Agriculture Natural Resources Conservation Services (NRCS), mandated to help reduce soil erosion and damages caused by floods and other natural disasters (10). Although neither agency had guidelines or a mandate to analyze suitability of a relocation site, both agencies conducted a series of studies regarding alternative relocation sites for Shishmaref.

In 2004 the Shishmaref Erosion and Relocation Coalition, which later dissolved as an organization, chose Tin Creek as the community's preferred relocation site. Between 2004 and 2008 the NRCS, USACE, and Alaska DOT conducted approximately six separate studies to evaluate Tin Creek's suitability as a relocation site (33). The DOT determined that the site was unsuitable because of the presence of ice-rich permafrost that could thaw as a result of climate warming and create future problems for community habitability (33). In June 2009, the City of Shishmaref received a grant through the ACCIMP to conduct a Shishmaref Site Selection Feasibility Study. As a consequence, the most recent relocation site analysis, conducted in 2010, recommended a relocation site 10 miles from the community, which may meet the community's need to be close to their traditional subsistence grounds and also meet government geophysical requirements (33). After geophysical tests are conducted to determine the site's suitability, the community will vote again to determine if this site also meets their needs (33). In 2011, the community created the Shishmaref Relocation Work Group to move the relocation effort forward. As in Kivalina, government agencies and the majority of community residents agree that relocation is the only adaptation strategy that will ensure the long-term resilience of the community, but the steps necessary to implement relocation, if the proposed site is approved, are unclear.

Newtok

Newtok, a Yup'ik Eskimo village, is located along the Ninglick River near the Bering Sea in western Alaska (35, 36). A combination of increased temperatures, thawing permafrost, and wave action has accelerated the erosion, causing the Ninglick River to move closer to the village (35). The State of Alaska spent about $1.5 million to control the erosion between 1983 and 1989 (26). Despite these efforts, erosion is projected to reach the school, the largest structure in the community, by about 2017 (35).

Six extreme weather events between 1989 and 2006 exacerbated these gradual biophysical changes. Five of these events precipitated FEMA disaster declarations (37). FEMA declared three disasters between October 2004 and May 2006 alone (37). These three storms accelerated the erosion and repeatedly "flooded the village water supply, caused raw sewage to be spread throughout the community, displaced residents from homes, destroyed subsistence food storage, and shut down essential utilities" (35). Public infrastructure that was significantly damaged or destroyed included the village landfill, barge ramp, sewage-treatment facility, and fuel storage facilities (26). The barge landing, which allows for most delivery of supplies and heating fuel, no longer exists, creating a fuel crisis. Salt water is affecting the potable water (26).

Newtok inhabitants voted three times, most recently in August 2003, to relocate to Nelson Island, 9 miles from Newtok (35). Newtok obtained title to their preferred relocation site, which they named Mertarvik, through a land-exchange agreement negotiated with the US Fish and Wildlife Service in 2003 (35). No infrastructure existed at the relocation site. In 2006, Newtok community residents built three houses at Mertarvik, with funding received by the Newtok Traditional Council. In 2009, construction of pioneer infrastructure, including a multipurpose evacuation center and barge landing, began at the relocation site through the work of the Newtok Planning Group.

Newtok Planning Group

The Newtok Planning Group is an informal boundary organization that emerged in May 2006 from an ad hoc series of meetings, when state and federal agencies realized that Newtok was serious about its relocation because it had chosen its relocation site, acquired legal title, assured geophysical stability, and constructed three homes (10, 17). No similar planning group was implemented to respond to the relocation efforts of Kivalina and Shishmaref.

The Newtok Planning Group is unique in Alaska in its multidisciplinary and multijurisdictional structure. The group consists of about 25 state, federal, and tribal governmental and nongovernmental agencies that all voluntarily collaborate to facilitate Newtok's relocation. The Alaska Department of Commerce, Community, and Economic Development (DCCED) is the lead coordinating Alaska state agency for the Newtok Planning Group, but no federal agency has authority to coordinate federal efforts for Newtok's relocation (10, 17). From the Newtok Planning Group's inception, the Newtok Traditional Council has led the relocation effort, ensuring that local needs and goals guide the process.

As is typical of boundary organizations, no state or federal statutes or regulations govern or guide the work of the Newtok Planning Group (34). Agency representatives had to educate each other about the laws, funding options, and limitations of each agency to identify and coordinate funding, including sharing equipment costs and coordinating its use (34, 38). State funding to build public infrastructure, such as schools and air landing strips, is extremely competitive. With no population permanently residing at the relocation site, Newtok has not yet been unable to secure funds to build this critical infrastructure (26).

Initial planning efforts focused on the design and construction of pioneer infrastructure consisting of an emergency evacuation center/community center, barge landing, and an access road that connects these two structures. Seven different federal, state and tribal entities are involved with the construction and funding of these facilities, but no agency is authorized with overall supervision of the project, which has caused delays (17). Construction of the evacuation center was not yet complete as of 2012.

Meeting the requirements of the National Environmental Protection Act, which requires environmental impact assessments of federally funded construction projects, has been a significant impediment to progress (17). The National Environmental Protection Act requires designation of a federal lead agency, but the Stafford Act and other legislation provide no federal agency with authority to take a lead role in community relocation (10). These statutory impediments to Newtok's relocation will affect all Alaskan communities seeking to relocate.

In summary, although Newtok has worked for approximately a generation (19 y) to relocate, with substantial supporting efforts from numerous government agencies, statutory and institutional barriers have caused significant delays of the relocation process. In addition, there are no mechanisms in place to ensure that the extensive intergovernmental learning and collaboration that has occurred in designing Newtok's relocation will assist with the relocation of Kivalina, Shishmaref, or other Alaskan communities.

DISCUSSION

Governance Limitations to Community Relocation in Alaska

In Alaska, the lack of an overarching institutional relocation framework has caused the relocation of Kivalina, Shishmaref, and Newtok to proceed in an ad hoc manner. Each community took a somewhat different approach to their relocation planning process. Newtok began a relocation planning process with the Alaska DCCED, whereas Kivalina and Shishmaref worked primarily with federal agencies, including the USACE. Kivalina attempted to use legal challenges to fund initial infrastructure, whereas Newtok engaged a complex group of agencies, some of which were able to access funds not specifically designated for relocation. Communities also differed in local governance structure. Newtok has only one governing body (the Newtok Traditional Council); Shishmaref formed a working group comprised of elders and tribal and city government representatives; and Kivalina worked through its two local governing bodies, the city government, which is a political subdivision of the State of Alaska, and the tribal council, which has a government-to-government relationship with the federal government of the United States (10).

The relocation site chosen by each community played an instrumental role in the willingness of state and federal government agencies to assist with relocation. The Immediate Action Workgroup recognized that, government needs to: "[c] reate a process/recipe to identify suitable relocation sites to ensure an efficient and successful outcome. Kivalina's experience is a reflection of the downsides of not having an effective process in place" (39). This process has not yet been established. Newtok chose a relocation site that was not subject to permafrost thaw and had a good water source. Both Kivalina and Shishmaref initially chose culturally appropriate relocation sites that were later opposed by federal and state government entities because of concerns with thawing permafrost. Kivalina eventually found a relocation site that meets government criteria for site suitability and is slowly moving toward relocation. The consulting firm hired by the City of Shishmaref recommended evaluation of a relocation site not previously considered by Shishmaref and suggested that additional geotechnical studies be performed to ensure the site's suitability for relocation. The absence of clear guidelines and criteria for site selection or funding for geotechnical evaluation delayed relocation efforts in Kivalina and Shishmaref, causing distrust and frustration with state and federal government authorities (24, 39).

Finally, consensus by the three communities and state and federal agencies that relocation was essential created barriers to repairing and maintaining storm-damaged infrastructure in the current locations. The statutory restrictions of the National Flood Insurance Program prevent government agencies from using funds to repair seriously deteriorated infrastructure because of their location in flood-prone areas unless the structures can be protected (30, 34, 37). For example, the design of a solid waste master plan in Newtok, Shishmaref, and Kivalina has been deferred because of each community's decision to relocate and the government's reluctance to build new infrastructure in an existing floodplain (10, 26, 27). As a result, "honey buckets," 5-gal buckets with plastic bag liners, are used in most homes instead of plumbing and sewage disposal (24,26, 27). A 2006 public health assessment found that sanitation conditions in Newtok were "grossly inadequate for public health protection" (26). Between 1994 and 2004, 29% of Newtok's children were hospitalized with lower respiratory tract infections (17). Destruction of Newtok's barge landing by storms raised the cost of essential supplies and infrastructure repair. In summary, the communities have been unable to relocate, but it is unsafe and unhealthy to remain where they are.

Strategies for Adapting Governance to Address Climate Change

Climate-induced population displacement requires a governance framework that can dynamically respond to communities faced with accelerating biophysical changes caused by increased temperatures. Adaptive governance, in this context, means that institutions need a range of options, including postdisaster recovery, protection in place (seawall/shoreline protection), hazard mitigation, and relocation, to respond to the humanitarian needs of communities.

Here we summarize a set of general strategy elements that emerge from relocation efforts by Alaskan communities and from other climate-change adaptation efforts. None of these strategy elements is essential or by itself guarantees success, but together they provide a toolkit for potentially successful adaptation to climate change. The toolkit is designed to create a multidisciplinary and multilevel assessment of climate-related risks that fosters leadership and integrates an iterative learning process to develop adaptation strategies (40, 41).

Identify Current Climate-Related Risks and Vulnerabilities and Project their Future Changes

Key components of governance of climate change adaptation are the capacity to monitor local social-ecological processes and implement a dynamic and locally informed institutional response (15, 42).

Kivalina, Shishmaref, and Newtok each documented the occurrence and damage from severe winter storms and accelerating rates of erosion that increasingly threatened lives and property. These assessments were confirmed by multiple agency reports. Global and Alaskan regional climate models project that severe winter storms will increasingly occur during ice-free conditions and that their erosional impact will be amplified by continued loss of protective sea ice (8, 9). The integration of local assessments with regional and national assessments can foster multilevel collaboration and well-structured dialogue among scientists, community leaders, and government representatives to develop adaptation strategies that minimize the societal risks of these climate changes (42, 43). For example, in our case studies, the communities participated in identifying climate-related risks by gathering data and making decisions about appropriate institutional responses to the hazard.

Adapt to Current Climate Extremes through known Adaptations and Adapt to novel impacts by exploring outside-the-box adaptation strategies.

Through funding for disaster relief, federal and state agencies spent about $32 million on erosion control projects intended to reduce erosion and risks to life and property, and projected in 2004 that Shishmaref alone would require an additional $90 million for infrastructure upgrades and erosion protection measures within 15 y (44). Alternatively, these funds could be used for relocation, which residents and agencies responsible for erosion and flood control concurred was the only viable adaptation option. However, as described above, there is no funding or governance mechanism to implement this adaptation. Without an institutional framework to identify the steps a community must take to begin a relocation process, communities will be caught in a maze of conflicting agency regulations, and relocation will proceed in an uncoordinated and ad hoc manner (26, 30, 34). Policy changes, which include the creation of an adaptive governance framework that can dynamically respond from protection in place to community relocation, are required for substantive progress toward relocation.

Integrate Ecological Integrity and Societal Well-being

Newtok's selection of a relocation site met the needs of both biophysical integrity (no high-ice-content permafrost and not highly susceptible to long-term coastal erosion or sea-level rise) and cultural integrity (continued opportunities for community cohesion and subsistence hunting activities). By including biophysical, cultural, and socioeconomic criteria in relocation planning, the relocation plan received widespread support from both community residents and government agencies seeking to assist with community relocation. Agency opposition to the relocation sites proposed by Kivalina and Shishmaref on the grounds of permafrost instability was a key impediment to relocation progress by those communities.

Integrate Climate-change Adaptation with other Societal Goals

Although community relocation is the most urgent challenge facing our three Alaskan communities, restrictions on repairing or upgrading current infrastructure create other hardships, such as high heating costs because of poor insulation, public health risks from inadequate sewage treatment, undependable fuel supply because of degraded barge-landing facilities, and high-maintenance, expensive, and inadequate water treatment, as observed in all three of our study communities. Community relocation provides an opportunity to address these multiple societal issues to foster long-term sustainability in the process of relocating communities. Mainstreaming of climate-change policies with other agency mandates increases the likelihood of efficient implementation (45, 46) and of accounting for the interactions between climate-induced impacts and other stressors (43).

Bridge among Formal Organizations to Facilitate Communication, Collaboration, and Learning

The Newtok Planning Group is an informal bridging organization that has worked intensively for 7 y to develop a relocation strategy despite the lack of any official relocation mandate for participation in the group. The collaboration that occurred created innovative solutions that were less likely to have emerged through formal channels. It remains to be seen whether the social capital thus created will contribute to relocation efforts of other villages. In general, bridging organizations and informal networks create new spaces where learning can occur and which are less constrained by the formal mandates of participating groups (47, 48). Bridging organizations may be particularly important in devising novel adaptation options or governance structures to improve the fit with the new conditions resulting from climatic and other global changes, for example the seasonally ice-free conditions in a warming Arctic Ocean (49).

Seek Interdisciplinary, Multisector Engagement that Fosters Local Leadership and Engages Local Governing Institutions in Identifying Potential Solutions

The breadth of stakeholder engagement by tribes, state and federal agencies, and nongovernment organizations in the Newtok Planning Group contributed to its success by reducing the likelihood of each agency and stakeholder group pursuing a separate and partially incompatible agenda (silos). Power-sharing and joint decision-making allowed learning to occur and created trust among participating groups spanning tribal, state, and federal entities (15). Leadership of the Newtok Planning Group by the Newtok Traditional Council ensured that solutions were place-based, local in scale, and understood and accepted by community residents. State and federal agencies along with nonprofit organizations, which have access to resources, geotechnical equipment to assess relocation sites, and expertise to build infrastructure, provided technical assistance to facilitate the community relocation. The Newtok Planning Group's collaborative governance structure, which recognized the need to address housing, transportation, and utilities as essential components of an integrated relocation strategy, has been essential in moving Newtok's relocation effort forward. Similarly, comprehensive multisector planning has been critical for complex adaptation planning at city, state, and national levels and is an important strategy to reduce and manage risk to climate extremes and disasters (41, 45).

Policy Implications

Our analysis suggests that climigration, as an effective adaptation strategy to climate change, requires a combination of local leadership to identify climate threats and potential solutions, elimination of higher-scale (e.g., state and national) institutional barriers that prevent effective local adaptation, and governance of climate-change adaptation that fosters innovation and efficient communication across these scales. Specifically, in Alaska, adaptation requires institutions to respond dynamically to accelerating climate-change impacts and prepare for a continuum of potential responses that include postdisaster recovery, protection in place (e.g., seawall and shoreline protection), hazard mitigation, and relocation. We therefore recommend the following:

- Amendment of federal policies such as the Stafford Act to include gradual and recurring climate-induced biophysical processes, such as erosion, would allow the President to declare such circumstances a disaster and release federal funds for predisaster hazard mitigation (42 U.S.C. § 5122) and planning as a response to climate change.

- Change in federal and state statutes to specifically permit federal disaster relief funding to be used and federal agencies to participate in building new infrastructure and relocating an entire community to a relocation site when durable adaptation is impossible in the current location.

- Creation of a relocation institutional framework to authorize government agencies to provide relocation technical assistance and funding, outline specific steps communities must take to begin a relocation planning process, (including the identification of site suitability criteria), and remove statutory barriers that impede relocation.

These amendments would allow Alaska Native villages and other communities threatened by climate-induced ecological changes to shift seamlessly from a disaster recovery to community relocation. The creation of this framework would avoid repeated humanitarian crises when communities are faced with chronic extreme weather events that accelerate biophysical change.

CONCLUSIONS

Climate-induced biophysical change threatens the lives, livelihoods, homes, health, and basic subsistence of many human populations. Governments and insurance companies may not be able to sustain the cost of rebuilding infrastructure repeatedly damaged or destroyed by these changes. Relocation may be the best adaptation response if the community's current location is uninhabitable, or relocation reduces vulnerability to future climate-induced ecological threats. We have outlined an adaptive governance framework that can respond to rapid directional environmental change involving extreme weather events to foster resilience in the face of these changes. Testing this framework for community relocation in Alaska provides an opportunity to learn and adaptively design institutional frameworks for a broader range of climate-change impacts in the United States and globally.

METHODS

To understand the community relocations occurring in Alaska, we conducted a case study of the relocation process in Kivalina, Shishmaref, and Newtok. Data-gathering tools used to collect evidence included surveys, interviews, participatory observation, and the study of organizational documents of the Newtok Planning Group, the Shishmaref Erosion and Relocation Coalition, and the Alaska Sub-Cabinet on Climate Change Immediate Action Workgroup. Archival document review included review of erosion assessments conducted by the USACE, results of the Newtok Housing Survey, community relocation lay-out documents, and geotechnical documents for each community, community relocation reports, and federal government relocation, erosion, and climate-change reports.

R.B. and F.S.C. participated in ~45 and 10 meetings, respectively, occurring on three different governance levels since 2007. These included meetings conducted by the Newtok Planning Group and the Immediate Action Workgroup and the Adaptation Advisory Group created by the Subcabinet on Climate Change.

ACKNOWLEDGMENTS

R.B. thanks Sally Russell Cox and the members of the Newtok Planning Group, the Newtok Traditional Council, and the Immediate Action Workgroup for allowing her to observe their numerous meetings. We thank Dennis Dixon, Sarah Trainor, Gary Kofinas, and Peter Schweitzer for constructive feedback. This study was supported by the National Science Foundation (Alaska Experimental Program to Stimulate Competitive Research and the Integrative Graduate Education and Research Training, Resilience and Adaption Program Award 0654441).

REFERENCES

1.   IPCC. Impacts, Adaptation and Vulnerability: Contribution of Working Group II to the Fourth Assessment Report of the Intergovernmental Panel on Climate Change. Cambridge: Cambridge Univ Press; 2007.

2.   CRED 2010. EM-DAT: The international disaster database, Available at http://www.emdat.be/natural-disasters-trends. (Centre for Research on the Epidemiology of Disasters, Brussels). Accessed March 21, 2013.

3.   Norwegian Refugee Council . Displacement Due to Natural Hazard-Induced Disasters: Global Estimates for 2009 and 2010. Geneva: Internal Displacement Monitoring Center, Norwegian Refugee Council; 2011.

4.   Buddemeier RW, Kleypas JA, Bronson RB. Coral Reefs and Global Climate Change: Potential Contributions of Climate Change to Stresses on Coral Reef Ecosystems. Arlington, VA: Pew Center on Global Climate Change; 2004.

5.   Nicholls RJ, Cazenave A. Sea-level rise and its impact on coastal zones. Science. 2010;328(5985):1517–1520.

6.   Woodworth PL. Have there been large recent sea level changes in the Maldive Islands? Global Planet Change. 2005;49(1–2):1–18.

7.   Ericson JP, Vörösmarty CJ, Dingman SL, Ward LG, Meybeck M. Effective sea-level rose and deltas: Causes of change and human dimension implications. Global Planet Change. 2006;50(1–2):63–82.

8. Mars JC, Houseknecht DW. Quantitative remote sensing study indicates doubling of coastal erosion rate in past 50 yr along a segment of the Arctic coast of Alaska. Geology. 2007;35(7):583–586.

9. Jones BM, et al. Increase in the rate and uniformity of coastline erosion in Arctic Alaska. Geophys Res Lett. 2009;36(3):L03503.

10. GAO . Alaska Native Villages: Limited Progress Has Been Made on Relocating Villages Threatened by Flooding and Erosion. Washington, DC: Government Accountability Office; 2009.

11. Bronen R. 2010. in *Environment, Forced Migration and Social Vulnerability*, eds Afifi T, Jäger J (Springer, Berlin), pp 87, 89.

12. GAO . Alaska Native Villages: Most Are Affected by Flooding and Erosion, but Few Qualify for Federal Assistance. Washington, DC: Government Accountability Office; 2003.

13. Feuer A (Nov. 4, 2012) Protecting the city, before next time.*New York Times*. Available athttp://www.nytimes.com/2012/11/04/nyregion/protecting-new-york-city-before-next-time.html?pagewanted=all&_r=0. Accessed March 21, 2013.

14. Moss ML, Shelhamer C. 2007. Cities, Communications and Catastrophe: Improving Robustness and Resiliency, The Stafford Act: Priorities For Reform. (Center For Catastrophe Preparedness and Response, New York University, New York, NY), Available at http://www.nyu.edu/ccpr/pubs/Report_StaffordActReform_MitchellMoss_10.03.07.pdf. Accessed March 21, 2013.

15. May B, Plummer R. Accommodating the challenges of climate change adaptation and governance in conventional risk management: Adaptive collaborative risk management (ACRM) Ecol Soc. 2011;16(1):47. Available at http://www.ecologyandsociety.org/vol16/iss11/art47/

16. FEMA . Hazard Mitigation Assistance Unified Guide.Washington, DC: Federal Emergency Management Agency, Department of Homeland Security; 2010.

17. Bronen R. Climate-induced community relocations: Creating an adaptive governance framework based in human rights doctrine. New York U Rev Law Social Change. 2011;35(2):356–406.

18. Ravens T, Jones BM, Zhang J, Arp CD, Schmutz JA. Process-based

coastal erosion modeling for Drew Point (North Slope, Alaska) J Waterw Port C-ASCE. 2012;138(3):122–130.

19. Stroeve J, Holland MM, Meier W, Scambos T, Serreze MC. Arctic sea ice decline: Faster than forecast. Geophys Res Lett.2007;34(9):L09501.

20. Francis OP, Panteleev GG, Atkinson DE. Ocean wave conditions in the Chukchi Sea from satellite and in situ observations. Geophys Res Lett. 2011;38(24):L24610.

21. ASCG . Newtok: Background For Relocation Report.Anchorage, AK: Arctic Slope Consulting Group; 2004.

22. USACE . Kivalina Relocation Master Plan. Anchorage, AK: US Army Corps of Engineers; 2006.

23. Gray G. Final Situation Assessment: Kivalina Consensus-Building Project. Juneau, AK: Glenn Gray and Associates; 2010.

24. Marino E. The long history of environmental migration: Assessing vulnerability construction and obstacles to successful relocation in Shishmaref, Alaska. Glob Environ Change.2012;22(2):374–381.

25. Berardi G. Schools, settlement and sanitation in Alaska native villages. Ethnohistory. 1999;46(2):329–359.

26. USACE . Project Fact Sheet. Anchorage, AK: US Army Corps of Engineers; 2008.

27. ANTHC . Climate change in Kivalina, Alaska. Anchorage, AK: Alaska Native Tribal Health Consortium Center for Climate and Health; 2011.

28. Darnell F. Education among the native peoples of Alaska. Polar Rec (Gr Brit) 1979;19(122):431–446.

29. Israel B (Nov. 9, 2011) Bering Sea Storm: Where did Alaska's 'epic' storm come from? *Christian Science Monitor.*

30. IAWG . Recommendations Report to the Governor's Subcabinet on Climate Change. Juneau, AK: Alaska SubCabinet on Climate Change, Immediate Action Workgroup; 2008. Available athttp://www.climatechange.alaska.gov/docs/iaw_rpt_17apr08.pdf. Accessed March 21, 2013.

31. D'Oro R (Dec. 31, 2011) Kivalina voters consider new school 7 miles away. *Anchorage Daily News*, p A1.

32. SERC . Shishmaref Strategic Relocation Plan. Shishmaref, AK: Shishmaref Erosion and Relocation Coalition; 2002.

33. BEESC . Shishmaref Relocation Plan Update Draft–Final Shishmaref, Alaska Shishmaref Erosion and Relocation Coalition and Kawerak Bristol Project #210029. Anchorage, AK: Bristol Environmental & Engineering Services Corporation; 2010.

34. IAWG . Recommendations Report to the Governor's Subcabinet on Climate Change. Juneau, AK: Alaska SubCabinet on Climate Change, Immediate Action Workgroup; 2009. Available athttp://www.climatechange.alaska.gov/docs/iaw_finalrpt_12mar09.pdf. Accessed March 21, 2013.

35. Cox S. An Overview of Erosion, Flooding, and Relocation Efforts in the Native Village of Newtok. Anchorage, AK: Alaska Department of Commerce, Community and Economic Development; 2007.

36. USACE . Revised Environmental Assessment: Finding of No Significant Impact: Newtok Evacuation Center: Mertarvik, Nelson Island, Alaska. Anchorage, AK: US. Army Corps of Engineers; 2008. Available athttp://www.commerce.state.ak.us/dca/planning/pub/Newtok_Evacuation_Center_EA_&_FONSI_July_08.pdf. Accessed March 21, 2013.

37. ASCG . Village Of Newtok, Local Hazards Mitigation Plan. Newtok, AK: ASCG Inc. of Alaska Bechtol Planning and Development; 2008. Available athttp://www.commerce.state.ak.us/dca/planning/pub/Newtok_HMP.pdf. Accessed March 21, 2013.

38. IAWG (2008) Meeting Summary, Jan. 18, 2008. (Alaska SubCabinet on Climate Change, Immediate Action Workgroup, Juneau, AK), Available athttp://www.climatechange.alaska.gov/docs/iaw_18jan08_sum.pdf. Accessed March 21, 2013.

39. IAWG 2008. Meeting Summary, March 4, 2008, (Alaska Sub-Cabinet on Climate Change, Immediate Action Workgroup, Juneau, AK)

40. Armitage D, Berkes F. In: Adaptive Co-Management: Collaboration, Learning, and Multi-Level Governance. Doubleday N, editor. Vancouver: Univ of British Columbia Press; 2007.

41. Lavell A, et al. 2012. in *Managing the Risks of Extreme Events and Disasters to Advance Climate Change Adaptation. A Special*

Report of Working Groups I and II of the Intergovernmental Panel on Climate Change (IPCC), eds Field CB, et al. (Cambridge Univ Press, Cambridge, UK, and New York, NY), pp 25–64.

42.   Dietz T, Ostrom E, Stern PC. The struggle to govern the commons. Science. 2003;302(5652):1907–1912.

43.   NRC . America's Climate Choices: Adapting to the Impacts of Climate Change. Washington, DC: National Academies; 2010.

44.   TetraTech . Shishmaref Partnership Shishmaref Relocation and Collocation Study Shishmaref, Alaska Preliminary Costs of Alternatives. Anchorage, AK: US Army Corps of Engineers; 2004.

45.   NYCPCC Climate change adaptation in New York City: Building a risk management response. Ann N Y Acad Sci.2010;1196:1–354.

46.   Commonwealth of Australia . Department of Climate Change Corporate Plan 2009–2010. Barton, Australia: Australian Government Department of Climate Change; 2009.

47.   Folke C, Hahn T, Olsson P, Norberg J. Adaptive governance of social-ecological systems. Annu Rev Environ Resour.2005;30:441–473.

48.   Margerum RD. A typology of collaboration efforts in environmental management. Environ Manage. 2008;41(4):487–500

49.   Berkman PA, Young OR. Science and government. Governance and environmental change in the Arctic Ocean. Science.2009;324(5925):339–340.

The Human Impact of Floods: A Historical Review Of Events 1980-2009 and Systematic Literature Review

Shannon Doocy,* Amy Daniels, Sarah Murray, and Thomas D. Kirsch*

*Shannon Doocy, Johns Hopkins Bloomberg School of Public Health, Baltimore, Maryland, United States.

ABSTRACT

Background. Floods are the most common natural disaster and the leading cause of natural disaster fatalities worldwide. Risk of catastrophic losses due to flooding is significant given deforestation and the increasing proximity of large populations to coastal areas, river basins and lakeshores. The objectives of this review were to describe the impact of flood events on human populations in terms of mortality, injury, and displacement and, to the extent possible, identify risk factors associated with these outcomes. This is one of five reviews on the human impact of natural disasters Methods. Data on the impact of floods were compiled using two methods, a historical review of flood events from 1980 to 2009 from multiple databases and a systematic literature review of publications ending in October 2012. Analysis included descriptive statistics, bivariate tests for associations and multinomial logistic regression of flood characteristics and mortality using Stata 11.0. Findings. There were 539,811 deaths (range: 510,941 to 568,680), 361,974 injuries and 2,821,895,005 people affected by floods between 1980 and 2009. Inconsistent reporting suggests this is an underestimate, particularly in terms of the injured and affected populations. The primary cause of flood-related mortality is drowning; in developed countries being in a motor-vehicle and male gender are associated with increased mortality, whereas female gender may be linked to higher mortality in low-income countries. Conclusions. Expanded monitoring of floods, improved mitigation measures, and effective communication with civil authorities and vulnerable populations has the potential to reduce loss of life in future flood events.

INTRODUCTION

Floods are the leading cause of natural disaster deaths worldwide and were responsible for 6.8 million deaths in the 20th century. Asia is the most flood-affected region, accounting for nearly 50% of flood-related fatalities in the last quarter of the 20th century [1,2,3]. The Center for Research on the Epidemiology of Disasters (CRED) defines a flood as "a significant rise of water level in a stream, lake, reservoir or coastal region" [4]. More colloquially, flooding is the "presence of water in areas that are usually dry" [1]. The events and factors that precipitate flood events are diverse, multifaceted, and interrelated. Weather factors include heavy or sustained precipitation, snowmelts, or storm surges from cyclones whereas important human factors include structural failures of dams and levies, alteration of absorptive land cover with impervious surfaces and inadequate drainage systems. Geographic regions such as coastal areas, river basins and lakeshores are particularly at risk from storms or cyclones that generate high winds and storm surge [5]. Environmental/ physical land features including soil type, the presence of vegetation, and other drainage basin characteristics also influence flood outcomes [6]. Floods transpire on varying timelines, ranging from flash floods with little warning to those that evolve over days or weeks (riverine). Flash floods, characterized by high-velocity flows and short warning times have the highest average mortality rates per event and are responsible for the majority of flood deaths in developed countries [1,3,7]. In contrast, riverine floods which are caused by gradual accumulation of heavy rainfall are less likely to cause mortality because of sufficient time for warning and evacuation. Occasionally floods are associated with secondary hazards such as mudslides in mountainous areas.

Recent accelerations in population growth and changes in land use patterns have increased human vulnerability to floods. Harmful impacts of floods include direct mortality and morbidity and indirect displacement and widespread damage of crops, infrastructure and property. Immediate causes of death in floods include drowning and trauma or injury [1,8]. Over an extended time period, there may also be increased mortality due to infectious disease [1,9,10,11]. The risks posed by future flood events are significant given population growth, proximities of populations to coastlines, expanded development of coastal areas and flood plains, environmental degradation and climate change [12]. The objectives of this review were to describe the impact of floods on the human population, in terms of mortality, injury, and displacement and to identify risk factors associated with these outcomes. This is one of five reviews on the human impact of natural disasters, the others being volcanoes, cyclones, tsunamis, and earthquakes.

METHODS

Data on the impact of flood events were compiled using two methods, a historical review of flood events and a systematic literature review for publications relating to the human impacts of flooding with a focus on mortality, injury, and displacement.

Historical Event Review

A historical database of significant floods occurring from 1980 to 2009 was created from publicly available data. Multiple data sources were sought to ensure a complete listing of events, to allow for both human and geophysical factors to be included, and to facilitate cross checking of information between sources. The two primary data sources were CRED International Disaster Database (EM-DAT) [4] and the Dartmouth Flood Observatory (DFO) Global Archive of Large Flood Events database [13]. For inclusion in the EM-DAT database, one or more of the following criteria must be fulfilled: 10 or more people killed or injured; 100 people affected; declaration of a state of emergency; or a call for international assistance. The DFO database provides a comprehensive list of flood events recorded by news, governmental, instrumental, and remote sensing sources from 1985 to 2009. Inclusion criteria are: significant damage to structures or agriculture, long intervals since the last similar event, or fatalities. Flooding specifically related to hurricane storm surge and tsunamis were excluded.

Event lists from both databases were downloaded in July 2007 and merged to create a single database; the database was updated in August 2009. The EM-DAT and DFO databases included 2,678 and 2,910 events, reported, respectively, between 1980 and 2009. Both EM-DAT and DFO reported the date and location of the event, the affected region and the number dead. In addition, the number affected, homeless, and total affected (sum of injured, homeless, and affected) were reported by EM-DAT. DFO also reported the number displaced, duration of the event (days), and 'flood magnitude.' Flood magnitude is a composite score of flood severity developed by DFO that encompasses damage level, recurrence interval, duration of the flood in days and the area affected [13]. For flood impacts reported by EM-DAT, zeroes were treated as missing values because they were used as placeholders and their inclusion in the analysis could contribute to the under estimation of tsunami impacts. The final list included 2,678 events reported by EM-DAT and 2,910 reported by DFO; 1,496 events were reported by both sources yielding a total of 4,093 flood events affecting human populations.

To assess risk factors for flood-related mortality the following categories were used: no deaths (0 deaths), low (1-9 deaths), medium (10- 49 deaths) and high (≥50 deaths). Bivariate tests for associations between flood mortality and the following characteristics were performed using χ^2 (categorical measures) and ANOVA (continuous measures): decade, region (defined by the World Health Organization (WHO)), income level (World Bank), gross domestic product (GDP), GINI (measure of income inequality), and flood magnitude. All covariates, with the exception of GINI, which was not strongly associated with flood mortality in adjusted analyses, and GDP, which was highly correlated with per capita World Bank income level, were included in the final multinomial logistic regression model to assess the relative risk of mortality at a given level as compared to events with no deaths. All analyses were performed using Stata Statistical Software, Version 11.0 [14].

Systematic Literature Review

Key word searches in MEDLINE (Ovid Technologies, humans), EMBASE (Elsevier, B.V., humans), SCOPUS (Elsevier B.V., humans), and Web of Knowledge, Web of Science (Thomson Reuters) were performed to identify articles published in July 2007 or earlier that described natural hazards and their impact on human populations. One search was done for all the five natural hazards described in this set of papers. This paper describes the results for cyclones. The systematic review is reported according to the PRISMA guidelines. Key words used to search for natural hazards included *natural hazard(s), natural disaster(s), volcano(s), volcanic, volcanic eruption, seismic event, earthquake(s), cyclone(s), typhoon(s), hurricane(s), tropical storm(s), flood(s), flooding, mudslide(s), tsunami(s), and tidal wave(s).* Key words included for impact on human populations were *affected, damage(d), injury, injuries, injured, displaced, displacement, refugees, homeless, wounded, wound(s), death(s), mortality, casualty, casualties, killed, died, fatality, fatalities* and had to be used in either the title, abstract or as a subject heading/key word. The search resulted in 2,747 articles from MEDLINE, 3,763 articles from EMBASE, 5,219 articles from SCOPUS, and 2,285 articles from ISI Web of Knowledge. Results from the four databases were combined and duplicates were excluded to yield a total of 9,958 articles.

A multi-stage screening process was used. First, title screening was performed to identify articles that were unrelated to natural disasters or human populations. Each title was screened by two independent reviewers and was retained if either or both reviewers established that inclusion criteria were met. To ensure consistent interpretation of inclusion criteria, percent agreement was assessed across reviewers for a small sample of articles, and title screening began after 80% agreement on inclusion was achieved. A total of 4,873 articles were retained for abstract review. Articles that met one or more of the following criteria were excluded in the abstract screening: language other than English; editorial or opinion letter without research-based findings; related to environmental vulnerability or hazard impact but not human populations; individual case report/study; focus on impact/ perceptions of responders; and not related to human or environmental vulnerabilities or impacts of hazards. As with the title screening, 80% overall agreement between reviewers was needed before abstract screening started. Each abstract was screened by two independent reviewers and was retained if either or both established that inclusion criteria were met. Included abstracts were coded for event type, timeframe, region, subject of focus, and vulnerable population focus. A total of 3,687 articles were retained for full article review. Articles discussing the impacts of natural disasters on human populations in terms of mortality, injury, and displacement were prioritized for review. A total of 119 articles on flood events meeting the criteria were retained for full review. Upon full review, 27 articles were retained including 17 that underwent standard data abstraction and 11 that were identified as review articles (Figure 1).

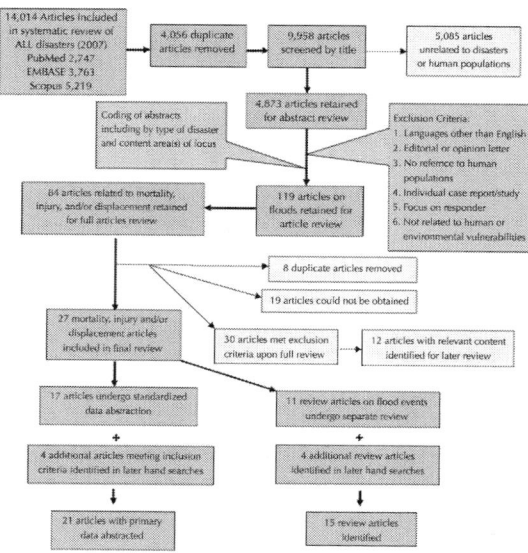

Figure 1: Overview of the systematic literature review process for floods.

Following the systematic review, a search was conducted to identify relevant articles published after the initial search up to October 2012. This search identified seven additional articles, including three articles with primary data that underwent full review and four review articles. Summaries of abstracted (n=21) and review articles (n=15) are presented in Tables 1 and 2, respectively.

Table 1: Articles included in the flood systematic literature review relating to mortality and injury* (abstracted, N=21)**

Article	Event	Summary	Mortality (n=15)	Injury and Morbidity (n=11)
Janerich, 1981[53]	Hurricane Agnes, 1972, New York, US	Epidemiologic investigation of cancer cases in rural town	Not reported	4 leukemia and lymphoma cases investigated; no increased risk due to flood/ environmental hazards identified
Duclos,1991[16]	October 1988,Nimes, France	Surveillance and household survey (n=108) to assess flood health effects	9 drowning deaths reported including two individuals attempting rescues; no risk factors reported	Injuries from surveillance (n=18) included: 3 severe, 3 near drowning, 2 hypothermia, and 10 minor injuries; 6% of 228 survey participants reported minor injuries
Siddique, 1991[17]	Mid-1988, Bangladesh	Record review of health facilities and verbal autopsy	9 of 154 (6%) deaths were directly due to flooding	5% (2,367/46,470) of patients had infected injuries
CDC, 1993[18]	Mid- 1993, Missouri, US	Public health surveillance and medical record review	27 deaths including 21 (78%) direct (drowning); 67% (n=18) of deceased were male	Not reported
CDC, 1993[34]	Summer 1993,Missouri, US	Surveillance of flood- related injuries and illnesses reported at hospitals	Not reported	524 flood-related conditions: 250 injuries (48%) and 233 (45%) illnesses; common injuries were sprains/strains (34%), lacerations (24%), abrasions/ contusions (11%)

CDC, 1994[19]	July, 1994,Georgia, US	Record review of flood-related deaths	28 deaths, 96% (n=27) due to drowning; at risk groups were males (71%), adults (86%), and car related (71%)	Not reported
Staes, 1994[20]	Jan 1992,Puerto Rico, US	Descriptive and case-control study of flood mortality	23 deaths; 22 (96%) drowning and 1 (4%) carbon monoxide poisoning; motor vehicles as risk factor	Not reported
Grigg, 1999[28]	July 1997,Colorado, US	Descriptive/ historical account	5 deaths reported; 80% were trailer park residents	54 injuries reported; no additional information reported
CDC, 2000[21]	Oct 1998, Texas, US	Public health surveillance and medical record review	31 deaths mostly from drowning (n=24, 77%) and trauma (n=3, 10%); most were male and car related	Not reported
Rashid, 2000[22]	1998, Dhaka Bangladesh	Qualitative survey	918 officially reported flood deaths; qualitative study observed 1200 deaths of which 2% were drownings	Not reported
Ogden, 2001[33]	May 1995,Louisiana, US	Surveillance and record review of disaster-area hospitals and patient visits	Not reported	1855 post-flood injuries, including musculoskeletal (n=791, 46%), lacerations (n=385, 21%), motor vehicle (n=142, 8%), falls (n=134, 7%), and other (n=296, 16%)
Yale, 2003 [23]	Sept 1999, North Carolina, US	Case-control study of vehicle crashes with drowning	ü 22 deaths reported; males and adults were disproportionately represented	Not reported
Cariappa, 2003[35]	July 2001,Orissa, India	Assessment of flood-related illness/ injury in care seekers	Not reported	13% (976/7450) of health facility visits due to injury; males and those 11-40yrs accounted for most injuries

Baxter, 2005[25]>	Jan & Feb1953, UK	Descriptive/ historical account	307 deaths due to drowning and exposure; elderly and coastal/ poor construction residents were most at risk	Not reported
Gerritsen, 2005[26]	Jan & Feb 1953, The Netherlands	Descriptive review / historical account	1836 deaths; no additional information reported	Not reported
Pradhan, 2007[24]	July 1993, Sarlahi District, Nepal	Household survey in flood affected areas	ü 302 deaths; CMR 7.3/1000; females and young children had greatest risk of death	Not reported
Spencer, 2007[27]	Summer 1977, Pennsylvania, US	Descriptive/ historical account	ü 78 deaths; no additional information reported	Not reported
Schnitzler, 2007[36]	August 2002, Saxony, Germany	Telephone survey of flood affected households	ü Not reported	55 (11.7%) of the survey population was injured; risk of injury was increased among those who came into contact with flood water (OR 17.8, 95% CI 17.8– 30.5).
Jonkman, 2009[29]	August 2005, New Orleans	Secondary data analysis of characteristics associated with flood-related mortality following hurricane Katrina	ü Overall mortality percent among exposed was 1%. 853 deaths reported, including 51% male (n=432) and 49% (n=421) female. The majority (85%, 705/829) were among those > 51 yrs of age. In deaths where race was reported (n=819), 55% were African American, 40% white, and 2% other.	Not reported

Biswas, 2010[37]	Summer 2007, Bangladesh	Household survey of child injury in flood-affected areas	ü Not reported	>18% (n=117) children injured were during flood; injuries included 38% lacerations, 22% falls, 21% drowning, 8% road traffic, 6% burns, 5% animal bites.
Bich TH, 2011[54]	October and November 2008, Hanoii, Vietnam	Cross-sectional household survey	ü 2 deaths, no additional information reported	27 injuries, including 18 lacerations/contusions/cuts, 3 fractures, 1 trauma and 5 others. Causes of injuries included falls (16), near-drowning (1) and other (10).

Table Note:-

* Displacement is excluded from the table because no primary data on displacement was collected in only one study, Schnitzler, 2007.

** Additional articles included from the hand searches are Schniztler 2007, Jonkman 2009, Biswas 2010 and Bich 2011.

Table 2: Review articles identified by the systematic review relating to mortality, injury, and displacement in flood events (N=15)

Article	Summary	Key Findings
Statistical Bulletin 1974[55]	Review of tornado, flood and hurricane associated mortality in the US from 1965 to 1974	More than 1,200 flood deaths in the United States during the review period with a concentrated in a few large events. 14 major river systems were linked to flood deaths; damage can be mitigated through reforestation, construction of reservoirs and flood walls, diversion, and improved early warning and forecasting systems.

French et al., 1983[45]	Review of National Weather Service flash floods reports from 1969 to 1981 to assess mortality effects of warning systems	Floods were the primary cause of weather-related deaths. There were 1,185 deaths in 32 flash floods with an average of 37 deaths per flood; the highest mortality was associated with dams breaking after heavy rains. Mortality was greater earlier in the study period and twice as many deaths occurred in areas with inadequate warning systems. 93% of deaths were due to drowning, of which 42% were car related.
Avakyan 1999[56]	Review of global flood events from 1997 to 1999 using Dartmouth Flood Observatory data	Damage due to floods increased over time due to more development in flood-affected areas; mapping and regulation of flood hazards zones are necessary to mitigate damage. Globally Bangladesh is the most affected by floods. Number of events, victims, evacuees and damage are reported for each year.
Berz, 2000[39]	Review of the impacts of major floods in the last half of the 20th century and summary of significant floods from 1990 to 1998 from the Munich Re natural event loss database	Floods account for half of all natural disaster deaths; trend analysis suggests the frequency of and damages associated with floods have increased over time. Excluding storm surges, the three most deadly flood events from 1990 to 1998 were in India, Nepal and Bangladesh in 1998–4750 deaths, China in 1998–3656 deaths, and China in 1993-3300 deaths. Explanations for increased mortality include population growth, vulnerability of structures, and construction in flood-prone areas, flood protection system failures and changes in environmental conditions.
Beyhun, Altintas & Noji, 2005[31]	Review of the impact of flooding in Turkey from 1970 to 1996	624 floods recorded during study period, including 83 fatal events with 539 deaths. There was an association between deaths and material losses, close to half of flood events occurred in summer months, and 37% of deaths in the Black Sea region.
Guzzetti, 2005[57]	Review of flood and landslide related deaths, missing persons, injuries and homelessness in Italy from 1279 to 2002	50,593 people died, went missing, or were injured in 2,580 flood and landslide events and over 733,000 were displaced. Floods accounted for 38,242 deaths; fatal events were most frequent in the northern Alpine regions and mortality was highest in autumn. Floods were caused by high-intensity or prolonged rainfall, snow melt, overtopping or failure of levees, embankments, or dams, and reservoir mismanagement. Since World War II, landslide has exceeded flood mortality and is comparable to earthquake mortality.

Jonkman & Kelman, 2005[1]	Examination of the causes and circumstances of 247 flood disaster deaths across 13 flood events in Europe and the US	Two-thirds of deaths were due to drowning. Being male and engaging in high risk behavior during flood events were also linked to increased flood mortality. Findings with respect to age-related vulnerability were inconsistent. Authors call for standardization of data collection methodologies across regions and flood types to improve policies and strategies to reduce flood-related death.
Jonkman, 2005[3]	Review of mortality from river floods, flash floods and drainage problems from 1975 to 2002 using the CRED Database	Of all disaster types, floods affect the most people; there were 1816 events with 175,000 deaths and 2.2 billion affected from 1975-2002. The deadliest freshwater flood events were Venezuela (1999, 30,000 deaths), Afghanistan (1998, 6,345 deaths), and China (1980, 6,200 deaths). Flash floods resulted in the highest average mortality per event. Average mortality (# fatalities / # affected) was constant across continents while impact magnitude (#s of dead and affected) varied between continents.
Tarhule, 2005[32]	Review of newspaper accounts of rainfall and rain-induced flooding in the Sahel savanna zone of Niger from 1970 to 2000	53 articles reported 79 damaging rainfall and flood events in 47 communities in the Sahel of Niger during the study period; floods destroyed 5,580 houses, killed 18, left 27,289 homeless, and caused over $4 million in damages. Sahel residents attribute floods to five major causes: hydrologic, extreme/unseasonable rainfall, location of affected area, inadequate drainage, and poor construction; cumulative rainfall in the days preceding a heavy rain event is an important predictor of flooding.
Lastoria, 2006[58]	Review of flood deaths and socioeconomic impacts in Italy, 1951 to 2003	During study period, ~50% of the flood events resulted in an average of 5 deaths, and about ~10% had >100 deaths. Investigators recommend creating an integrated database to collect more information about flood events in Europe.
Llewellyn, 2006[44]	Review mortality, injury, illness and infectious disease associated with major, recent floods events	In the US, as much as 90% of natural disaster damage (excluding droughts) is caused by floods which cost $3.7 billion annually from 1988 to 1997. There were an average of 110 flood deaths/yr from Between 1940 to 1999, mostly in flash floods and automobile related. Most flood related injuries are mild, and predominantly consist of cuts, lacerations, puncture wounds, and strains/sprains to extremities.

Ahern, 2005[30]	Review of studies of global flood events and assessment of gaps in knowledge relative to reducing public health impact of flooding	Review of 212 epidemiologic studies with detailed findings reported for 36 studies. The majority of flood deaths were due to drowning; deaths due were diarrhea inconclusive though there is some evidence to support increased risk of fecal-oral disease, vector-borne disease and rodent-borne disease. There is a lack of data on frequency of non-fatal flood injury.
Ashley & Ashley, 2008[8]	Review of flood fatalities in the United States from 1959 to 2005	4,585 fatalities over a 47 year period were reported (97.6 deaths/year). No significant increase in flood mortality over time was observed. The majority of flood-related deaths were in flash floods and were motor-vehicle related (63%). Increased risk of flood-related death was observed in individuals ages 10-29 and >60 years.
Jonkman & Vrijling, 2008[49]	Review of mortality attributed to different flood types and presentation of new method for estimating flood related deaths in low-lying areas	Reports on 1883 coastal flood events between 1975 and 2002 resulting in 176,874 deaths and 2.27 billion affected. Mortality by event type was reported as follows: 70 from drainage floods, 392 from river floods and 234 from flash floods. Flood mortality was affected by severity of flood impacts and warning and evacuation. Primary determinants of flood-related death include: lack of warning, inability to reach shelter, building collapse, water depth, rapid rise in water level, water flow velocity, children, and elderly. Applies a new method for estimating loss of life due to floods based on flood characteristics and numbers exposed and mortality among exposed are introduced.
FitzGerald, 2010[50]	Review of flood fatalities in Australia from 1997 to 2008	Estimated 73 flood-related deaths reported from newspapers and historic accounts from 1997 to 2008 in Australia. Most fatalities occurred in the summer months. Drowning deaths were more likely among individuals between the 10-29 and >70 years of age. No difference decline in deaths over time reported. 49% of deaths were motor-vehicle related and 27% were attributed to high risk behavior.

RESULTS

Historical Event Review

Overall, an average of 131 (range 35-287) floods affected human populations annually with the majority (81%) occurred during or after the 1990s. Part of this increase can be explained by improved reporting and by the DFO reporting beginning in 1985. There was great variation in the number of events reported annually between EM-DAT (range 35-213) and DFO (42-235) (Figure 2). While the frequency of flood events increased gradually over time, their impacts on human populations in terms of mortality and affected populations varied greatly between years and were often concentrated around large-scale events (Figure 3). Using the WHO regions the Americas (AMRO) and Western Pacific (WPRO) regions experienced the most flooding events while the fewest were reported in Europe (EURO) (Figure 4). Deaths were overwhelmingly concentrated in South East Asia (SEARO), which accounted for 69% of global flood mortality, though both the Americas (AMRO) and Western Pacific (WPRO) had significant minorities of flood fatalities. The great majority of the flood affected population was in WPRO (59%) and SEARO (35%) of the global total. Overall, the human impacts of floods in Europe, Africa, and the Eastern Mediterranean regions were limited; together the regions accounted for no more than 8% of flood deaths and 4% flood affected populations, respectively. The overall impact of flooding on human populations is summarized in Table 3.

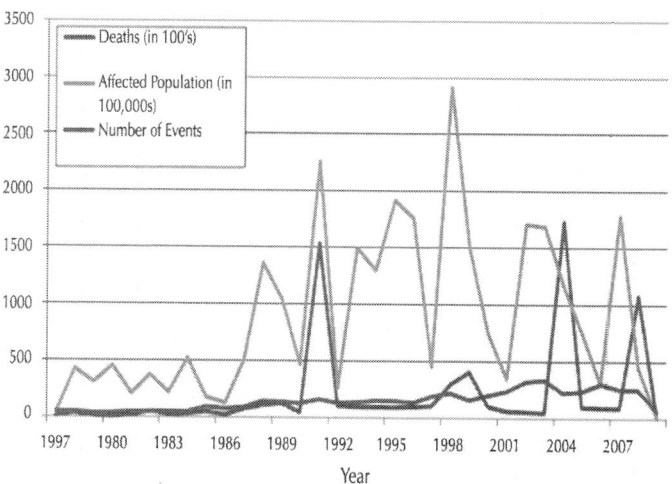

Figure 2: Reporting of flood events by source and year.

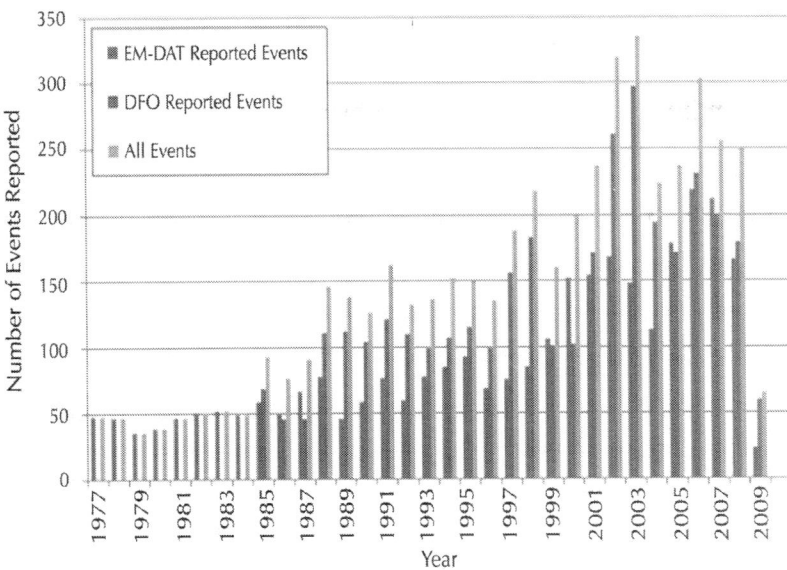

Figure 3: Flood events affecting human populations by year.

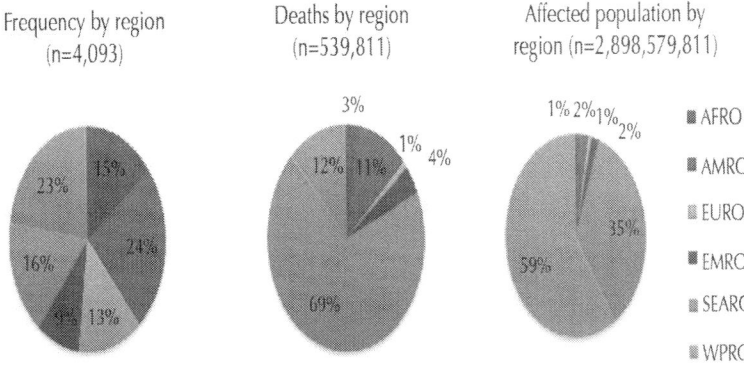

*Regions as defined by WHO; DFO reporting starts in 1985

Figure 4: Regional summary of flood events and their effects on human populations, 1980-2009*.

Table 3: Summary measures for the impact of floods on human populations, 1980-2009 (N=4,093)*

Reported Overall Impact of Flooding Events					
Human Consequence	**# of Events**		**Best Estimate**		**Range**
Deaths	4,093		539,811		510,941-568,680
Injuries	401		362,122		---
Homeless	611		4,580,522		---
Total Affected	2,632		2,898,579,881		---
Event Summary Statistics					
Human Consequence	**# of Events**		**Median**	**Mean**	**Range**
Deaths, all events	**3,960**	96.8%	**9**	**135**	**0-138,000**
Reported by EM-DAT	2,646	64.6%	10	74	0-30,000
Reported by DFO	2,732	66.75%	11	166	0-138,000
Events with deaths	**2,673**	65.3%	**11**	**146**	**1-138,000**
Reported by EM-DAT	2,146	52.4%	10	87	1-30,000
Reported by DFO	1,289	31.5%	13	178	1-138,000
Injured, all events	401	9.8%	12.5	904	0-249,378
Homeless, all events	611	14.9%	15	7,506	0-2,951,315
Total Affected, all events	2,632	64.3%	6,000	1,071,829	0-238,973,000

Table Note:

*Figures are based on the highest reported number of deaths or injuries in an event. Deaths were reported in 4,093 events. Homeless, injured, and total affected populations are reported only by EM-DAT, thus ranges are not presented for overall impact estimates.

Affected Population: An estimated 2.8 billion people were reported to be affected by flood events between 1980 and 2009, including nearly 4.6 million rendered homeless. However, these figures likely substantially underestimate the true impact of floods on human populations because estimates of the total affected population and the homeless population were reported in only 64.3% (n=2,632) and 14.9% (n=611) of events, respectively. The distribution of the number affected was highly skewed with mean and median affected populations of 1,071,829 and 6,000 per event, respectively, which indicates that the median affected population may better reflect the impact of a typical flood event.

Mortality and Injury: When mortality data from the two sources were combined, deaths were reported in 96.8% (n=3,960) of floods since 1980. This figure excludes 13.9% of floods where no information on mortality was reported; if no deaths are presumed and these events are included, deaths occurred in 65.3% (n=2,673) of floods. 539,811 deaths (range: 510,941-568,680) resulting from flood events were reported. For floods where mortality was reported, there was a median of 9 (mean=135; range 0-138,000) deaths per event when using the highest reported death toll. Mortality exceeded 10,000 in only 4 events and 100,000 in two. The two deadliest events occurred in Bangladesh (138,000 deaths in 1991) and Myanmar (100,000 deaths in 2008). Injuries were reported in 401 (9.8%) events, where a total of 361,974 injuries were documented. In events where injuries were reported, there was a median of 12.5 (mean=904: range 1-249,378) per flood event. To estimate the total number of injuries due to flood events, it was presumed that injuries would occur in events where deaths were reported. There were 2,673 floods with fatalities but only 401 (9.8%) with injuries reported. When the median and mean for injuries were applied to the remaining 3,077 events, it was estimated that between 38,463 and 2,717,681 additional unreported flood related injuries may have occurred between 1980 and 2009.

Bivariate associations between country-level characteristics and flood-related mortality from 1980 through 2009 are presented in Table 4. Findings suggests that the proportion of events with high mortality (>50 deaths) have decreased over time. Income level was also significantly associated with flood mortality, where for both low and lower-middle income countries, a greater proportion of events fell in the medium and high death categories as compared to higher income countries. Higher mortality events were concentrated in the South East Asian and Western Pacific regions.

Table 4: Flood event mortality characteristics, 1980-2009 (N = 4,093)

Characteristic	No deaths (n=706)	1-9 deaths (n=1,378)	10-49 deaths (n=1,223)	>50 deaths (n=785)	P-value
Decade, N (%)					
1980	121 (17%)	149 (11%)	212 (17%)	205 (26%)	<.001
1990	191 (27%)	418 (30%)	437 (35%)	317 (40%)	
2000	394 (55%)	811 (58)	574 (45%)	263 (33%)	
World Bank Development Level, N (%)					
Low income	172(24%)	263 (20%)	370 (30%)	365 (45%)	<.001
Lower Middle income	164 (23%)	395 (29%)	465 (38%)	328 (41%)	
Upper-middle income	142 (20%)	276 (21%)	219 (18%)	79 (10%)	
High Income	227 (32%)	408 (30%)	176 (14%)	33 (4%)	
World Health Organization Region, N (%)					
Africa	139 (20%)	228 (17%)	157 (13%)	73 (8%)	<.001
Americas	182 (26%)	387 (29%)	293 (24%)	122(15%)	
Eastern Mediterranean	46 (6%)	107 (8%)	147 (12%)	74 (9%)	
European	171 (23%)	246 (18%)	104 (9%)	26 (3%)	
South East Asian	47 (7%)	137 (10%)	229 (19%)	264 (33%)	
Western Pacific	124 (18%)	238 (18%)	299 (24%)	262 (32%)	
Gross Domestic Product, per capita, mean (SD), (n=4,089)	14,827 (18,077)	14,330 (17,710)	1,457(12,563)	3,325(6,518)	<.001
GINI,* mean (SD), (n=3,830)	40.2 (7.6)	41.0 (7.7)	41.7 (7.9)	41.3 (7.1)	0.004
Magnitude, mean (SD), (n=2911)**	4.8 (1.2)	4.9 (1.1)	5.3 (1.0)	6.0 (1.1)	<.001

Table Note:

*GINI coefficient scores for income distribution range from 0 to 100 with 0 representing a perfect equality and 100 perfect inequality.[59]

** Magnitude is a composite score of flood severity created by DFO that includes flood duration and affected area size, with the following categories: low magnitude,6.0. Flood magnitude is only available for events from 1985 onward.

Findings from the adjusted analyses (Table 5) modeling the relative risk of flood related mortality show that all predictors were significantly associated with flood mortality. The relative risk of medium- and high-level mortality events compared to events with no deaths significantly decreased over time. There was also a significant decreased relative risk of mortality in excess of 50 deaths for events in higher income countries compared with lower income country events. Additionally, as magnitude of a flood increased, so did the risk of having high mortality when adjusting for all other predictors. A flood rated as high magnitude as compared to one with low magnitude was associated with an increased relative risk of having high mortality as compared to no mortality (RR=13.20, 95% CI 8.25, 22.11). Caution should be taken when interpreting such findings, however, as magnitude estimates were missing for a large proportion of events, and missing magnitude was associated with the outcome in this study. Regional differences in reported mortality were also supported by the analysis. Higher mortality events were concentrated in the South East Asian and Western Pacific regions, compared to events occurring in the Americas (Southeast Asia RR=3.35, 95 CI: 2.21, 5.72; Western Pacific RR=2.38, 95 CI: 1.62, 3.34).

Table 5: Multinomial logistic regression results for mortality in flood events, 1980-2009 (N =4,093)*

Characteristic	1-9 deaths COR (95% CI)	P-value	10-49 deaths COR (95% CI)	P- value	>50 deaths COR (95% CI)	P-value
Decade						
1980	Reference		Reference		Reference	
1990	1.09 (0.87, 1.37)	.426	1.64 (1.29-2.07)	<.001	2.61 (1.99-3.42)	<.001
2000	0.86 (0.64, 1.15)	.313	1.85 (1.39-2.46)	<.001	4,46 (3.22-6.18)	<.001

World Health Organization Region							
AMRO	Reference		Reference		Reference	
AFRO	1.09 (0.76-1.55)	.0.62	0.58 (0.41-0.84)	.005	0.35 (0.22-0.56)	<.001
EURO	0.72 (0.54-0.96)	.024	0.45 (0.32-0.63)	<.001	0.31 (0.18-0.52)	<.001
EMRO	1.31 (0.83-2.06)	.240	1.49 (0.95-2.33)	.082	1.31 (0.78-2.21)	.3120
WPRO	0.80(0.59-1.09)	.165	1.22 (0.88-1.67)	.217	2.38(1.62-3.49)	<.001
SEARO	1.61(1.04-2.49)	.032	2.15 (1.40-3.29)	<.001	3.35 (2.21-5.72)	<.001
World Bank Income Level						
Low	Reference		Reference		Reference	
Lower middle	152 (1.06-1.92)	0.007	0.99 (0.74-1.34)	.992	0.59 (0.43-0.82)	0.002
Upper middle	1.56 (1.05-2.13)	0.014	0.90 (0.62-1.29)	.576	0.39 (0.24-0.61)	<.001
High	1.16 (0.86-1.71)	0.400	0.29 (0.20-0.42)	<.001	0.05 (0.03-0.08)	<.001
Flood Magnitude Category**						
Low	Reference		Reference		Reference	
Medium Low	1.03 (0.74, 1.44)	.859	1.47 (1.03, 2.10)	.035	1.52 (.95, 2.43)	.0878
Medium High	1.19 (0.85, 1.69)	.310	2.19 (1.50, 3.16)	<.001	3.87 (2.45, 6.10)	<.001
High	0.91 (0.62, 1.35)	.664	2.37 (1.58, 3.55)	<.001	13.20 (8.25, 21.11)	<.001
Missing	0.19 (0.15, 0.25)	<.001	0.32 (0.24, 0.43)	<.001	0.59 (0.40, 0.87)	.007

* Reference is "no deaths" for all categories (n=743) **see Table 4 notes for definition of flood magnitude

Systematic Literature Review

Mortality. Fourteen of the reviewed articles reported mortality data including ten that provided information on direct or indirect causes of mortality and/or risk factors for flood-related deaths (Table 6) [15,16,17,18,19,20,21,22,23,24,25,26,27,28] . Most articles provided some information about the distribution of deaths across population subgroups (i.e. gender, age) and/or an individual's location at the time of the event; seven of these ten articles reported on floods in the United States. Nearly all articles reporting cause of death cited drowning as the most frequent cause of death [1,15,18,19,20,22,29]. Cumulatively, drowning accounted for 75% of deaths; other causes of death included falls, electrocution, heart attack, hypothermia, trauma, snake bites, and carbon monoxide poisoning.

All studies in the United States examined mortality related to motor vehicles and found an increased risk of mortality among individuals in motor vehicles during the event, of all deaths 74% were motor vehicle related [17,18,19,20] . This compares to a motor vehicle related death rate of 63% in a recent review of US flood fatalities between 1959 and 2005[7]. Higher proportions of deaths among males (64%) were consistently observed in the United States, except for Puerto Rico where 57% (13/23) of flood related fatalities were female and hurricane Katrina where deaths evenly divided between the sexes (51% male, 49% female) [16,18,19,20,28] . In contrast, the one article describing flood mortality in the less developed country of Nepal found that females of all age groups faced increased mortality risk and 58% of all deaths were women [23] Other factors found to be associated with flood-related mortality included storm course/time storm hit landfall [19,22] summer months [17,30], low socioeconomic status [23], poor housing construction [,23,24,31] and timing of warning messages [19,22].

Table 6: Primary research articles describing flood related deaths and risk factors for flood mortality (N=10)

Article	County & Year	Flood Related Deaths			By Cause		By Sex		By Age	Vehle Related
		Total	Direct	Indirect	Drowning	Other Causes	Males	Female		
Duclos,1991[16]	France, 1988	9	9 (100)	0 (0%)	9 (100%)	0 (0%)	Not reported		Not reported	Not reported
CDC, 1993[18]	USA, 1993	27	21 (78%)	6 (22)	21 (78%)	2 (7%) electrocution2 (7%) vehicle accident 2 (7%) cardiac arrest	18 (6%)	9 (3%)	Average age = 38(range 9-88)	13 (48%)
CDC,1994[19]	USA, 1994	28	27 (96%)	1 (4%)	27 (96%)	1 (4%) other	20 (71%)	8 (29%)	Average age = 31(range 2-84)	20 (71%)
Staes,1994[20]	USA, 1992	23	22 (96%)	1 (4%)	22 (96%)	1 (4%) carbon monoxide poisoning	10 (43%)	13 (57%)	16 (70%) ≥ 16 yrs	20 (87%)
Grigg, 1999[28]	USA, 1997	5	5 (100%)	0 (0%)	Not reported		5 (100%)	0 (0%)	All adults	Not reported
CDC, 2000[21]	USA, 1998	31	29 (94%)	2 (6%)	24 (77%)	3 (10%) trauma1 (3%) hypothermia1 (3%) cardiac arres2 (6%) other	20 (65%)	11 (35%)	Median age = 38(range 2-83)	22 (71%)
Rashid, 2000[22]	Bangladesh, 1998	50*	Not reported		24 (48%)	21 (42%) electrocution 5 (10%) snake bites	Not reported		Children accounted for 92% (22/24) of drownings	Not reported

Yale, 2003[23]	USA, 1999	22	22 (100%)	0 (0%)	22 (100%)	0 (0%)	17 (77%)	5 (23%)	21 (95%) adults	22 (100%)
Pradhan, 2007[24]	Nepal, 1992	302	Not reported		Not reported		126 (42%)	176 (58%)	164 (54%) children138 (46%) adults	Not reported
Jonkman et al., 2009[29]	USA, 2005	853	Not reported		Not reported		432 (51%)	421 (49%)	705 (85%) older than 51 yrs, 60% over 65 yrs	Not reported
Totals		447	135 (93%)	10 (7%)	125 (75%)	42 (25%)	639 (50%)	643 (50%)	---	97 (74%)

*excludes 1150 deaths from diarrhea and other possibly deaths reported during the 4 month period surrounding the event

Injury and Displacement: Injury or morbidity data were reported in ten of the 18 included articles, of which nine provided information on injury type and/or risk factors[15,16,24,32,33,34,35,36,54] . The majority of flood-related injuries are minor. The two studies that captured a large number of injuries, both in the United States, found that musculoskeletal injuries were most common (46% and 34%), followed by lacerations (21% and 24%). Other flood-related injuries included abrasions and contusions, motor vehicle related injuries, and falls [33,34,54]. In less developed settings, increased incidence of snake bites and fires were also cited as causes of injury or death [2,36]. Among care seekers in flood-affected areas of Bangladesh 5.1% of wounds were infected. Another review suggested that the proportion of survivors requiring medical attention is less than 2% [2]. A distribution of injuries across population subgroups was reported by only one study in India which found that injuries were more common in males (67% vs. 33%), that the 11-40 year age group comprised 68% of the injured, and that those age 50 and above accounted for 18% of flood deaths [34]. Seven articles reported displacement or evacuation figures however none described risk factors associated with flood-related displacement[15,17,21,24,25,35,37] .

DISCUSSION

Main Findings

In the past 30 years approximately 2.8 billion people have been affected by floods with 4.5 million left homeless, at approximately 540,000 deaths and 360,000 injuries, excluding an estimated 38,000 to 2.7 million injuries that went unrecorded. While the mortality estimate presented in this study is consistent with the range of estimates presented in other studies [1,38], approximations of numbers injured and displaced are likely gross underestimates of the true values given the infrequency with which figures are reported. Floods events with high levels of mortality are relatively rare: despite their increasing frequency, there were only four events with >10,000 deaths and 58 events with >1000 deaths between 1977 and 2009. A slight decrease in the average number of fatalities per event was observed which is in keeping with broader natural disaster trends that show an increase in the size of the affected population and a decrease in the average number of deaths per event [4]. Higher numbers of fatalities were reported in flash floods than river floods, however, river floods affected larger populations and land areas [3,7]. Lower mortality rates in river floods can mostly be attributed to their slower onset allowing for longer time for warning and evacuation [3,39]. The widespread use of effective early warning methods for hydrological events has likely contributed declining flood mortality.

Findings from the historical event review are consistent with previous observations that flood mortality varies by region, economic development level, and the severity of the event [12,40]. The majority of flood-related deaths are concentrated in less developed and heavily populated countries, with Southeast Asia and the Western Pacific region experiencing the highest risk of flood-related deaths. Flood mortality rates are relatively similar across continents, but Asian floods kill and affect more people because they affect substantially larger areas with larger populations [3]. At the country level, lower GDP per capita was linked to higher mortality, which is in keeping with the established relationship between poverty and increased disaster risk [41]. Human and social vulnerabilities and inequalities, urbanization, population density, terrain and geo-physical characteristics and variation in the frequency and precipitating causes of floods by region are also factors that contribute flood risk levels [3,6,12,42]. Temporal changes and development trends have also contributed to changing influences of some of these factors over time [42]. Economic development increases the risk of disaster-related economic losses however improved emergency preparedness, response, and coping capacity may reduce disaster vulnerability[3]. That countries with greater resources are able to better predict and respond to impending flood events suggests that building systems and capacity to detect and respond to floods in less developed countries should be a priority [40].

Causes of and risks for flood-related mortality and injury identified in the systematic literature review are consistent with previous reviews on the human impact of flooding[1,29,43,44]. In comparison, a recent review of 13 flood events in Europe and the United States found that 68% of deaths were due to drowning, 12% trauma, 6% heart attack, 4% fire, 3% electrocution, 1% carbon monoxide poisoning, and 7% other/unknown [1]. Studies reporting the gender breakdown for flood-related deaths, most of which are accounts of flood events in the United States, consistently show a greater proportion of males as compared with female deaths. These observations are aligned with previous studies, including a review of flood events in Europe and the US which estimated that males account for 70% of flood related deaths [1,44,45,46]. While limited to only a few countries, these findings suggest there may be increased mortality risk for males in more developed settings and for females in less developed countries [23,47]. An increased risk of death in younger and older populations was also observed which is consistent with broader natural disaster mortality trends [7,45,46,48,49]. In Nepal, children had the highest crude mortality rates of all age groups and were nearly twice as likely to die in the flood as their same-sex parent [23]. However, recent reviews of age-specific risk for flood mortality have been inconclusive because attempts to aggregate data were hampered by high proportions of deaths where age is unreported [1]. While the prevailing notion is that women and children are more vulnerable in disasters [50], there is a paucity of research in less developed countries where the majority of flood deaths occur. Future research on the human impacts of floods should focus on these less developed settings, most notably Asia where flood deaths are concentrated, with the aim of identifying the most at-risk and vulnerable population sub-groups to better target early warning and preparedness efforts.

The ecological nature of the study of event characteristics did not allow for an examination of specific factors within a country or region that may be associated with increased mortality following a flood event. Population density in coastal regions, which are particularly vulnerable to flooding, is twice of the world's average population density and many of the world's coasts are becoming increasingly urbanized [51]. Currently, 50.6% of the world's population lives in urban settings; by 2050 this figure is projected to increase to 70% with the majority of urbanization occurring in less developed regions of Asia and Africa [52]. Unabated urbanization and land use changes, high concentrations of poor and marginalized populations, and a lack of regulations and preparedness efforts are factors that will likely contribute to an increasing impact of floods in the future [38]. From the natural hazard perspective, climate change is also likely to contribute to future increases in flooding. Increased frequency of intense rainfall, as a result of higher temperatures and intensified convection will likely lead to a rise in extreme rainfall events, more flash floods and urban flooding due to excessive storm water. Additionally, sea level rise and increasing storm frequency will lead to additional storm surges in coastal areas while seasonal changes, notably warmer winters, will contribute more broadly to increased precipitation and flood risk [38]. Together, changes in socioeconomic, demographic, physical terrain features and climatologic factors suggests that floods will become more frequent and have greater effects on human populations in the coming decades.

Given that flood losses are likely to increase in future years, increased attention to flood prevention and mitigation strategies is necessary. To date, early warning systems have been an effective mechanism for reducing the impact of floods [38], however, they are not ubiquitous and should be prioritized in less developed countries with large at-risk populations and high frequencies of flooding. It is important that messaging and targeted communication strategies accompany early warnings so that the population understands the impending risk and can respond appropriately. Many flood fatalities are associated with risk-taking behaviors, thus messages to avoid entering flood waters and to curtail risky activities in all stages of the event may be successful in reducing flood fatalities [1]. Additional, improved land use planning and regulation of development can mitigate flood impacts. Studies on the relationships between flood losses, natural hazard characteristics, and societal and demographic vulnerability factors can aid in informing and prioritizing flood prevention and mitigation strategies. Finally, comparisons of the effectiveness of different policies and mitigation strategies can inform future strategy and policy actions and ensure they are appropriate in specific contexts.

Limitations

The effects of flood events are the subject of gross approximations and aggregations that have a great deal of imprecision. The availability and quality of data has likely increased and improved over time and the use multiple data sources increased reporting. However, in many events deaths are unknown or unrecorded; for other outcomes such as injured and affected, reporting frequency is even lower which likely contributes to a substantial underestimation of the impacts of flood events on human populations. While available data is sufficient for a cursory analysis of global flood impacts and trends, improved reporting of flood outcomes, including the development of national systems capable of more accurately reporting mortality and injury would be beneficial.

Regarding the measures used in this study, our multivariable model included a broad classification of income level according to the World Bank, as opposed to GDP. While we believe GDP to be a more precise measure of wealth, it was nonetheless excluded in the analysis because we did not obtain GDP estimates that were time specific to each event. Inconsistencies and errors were common in data files from different sources, and in some cases inclusion criteria were not ideal for the purposes of this review, which created a challenge in reconciling event lists. For example, the 2004 Asian tsunami was classified as a flood by Dartmouth but not by EM-DAT; this event was ultimately removed from the data set, however, it represented the highest mortality event in the study period, which has potentially important implications for analysis. Consistent definitions and categorization of events across sources such as that initiated by EM-DAT in 2007 would be useful for streamlining future analysis and comparing the impacts of different types of flood events. Other principal limitations of the literature review are 1) that an in-depth quality analysis of all reviewed articles was not undertaken, and 2) the fact that only English language publications were included which likely contributed to incomplete coverage of studies published in other languages originating from low and middle income countries.

CONCLUSIONS

Interpretation of flood fatality data is challenging given the occurrence of occasional extreme events, temporal trends and the completeness and accuracy of available data. The continuing evolution of socio-demographic factors such as population growth, urbanization, land use change, and disaster warning systems and response capacities also influences trends. Between 1980 and 2009 there were an estimated 539,811 deaths (range 510,941 -568,584) and 361,974 injuries attributed to floods; a total of nearly 2.8 billion people were affected by floods during this timeframe. The primary cause of flood-related mortality was drowning. In developed countries being in a motor-vehicle at the time of a flood event and male gender were associated with increased mortality risk. Female gender may be linked to higher mortality risk in low-income countries. Both older and younger population sub-groups also face an increased mortality risk. The impact of floods on humans in terms of mortality, injury, and affected populations, presented here is a minimum estimate because information for many flood events is either unknown or unreported.

Data from the past quarter of a century suggest that floods have exacted a significant toll on the human population when compared to other natural disasters, particularly in terms of the size of affected populations. However, human vulnerability to floods is increasing, in large part due to population growth, urbanization, land use change, and climatological factors associated with an increase in extreme rainfall events. In the future, the frequency and impact of floods on human populations can be expected to increase. Additional attention to preparedness and mitigation strategies, particularly in less developed countries, where the majority of floods occur, and in Asia, a region disproportionately affected by floods, can lessen the impact of future flood events.

ACKNOWLEDGMENTS

We are grateful to Sarah Bernot, Dennis Brophy, Georgina Calderon, Erica Chapin, Joy Crook, Anna Dick, Shayna Dooling, Anjali Dotson, Charlotte Dolenz, Rachel Favero, Annie Fehrenbacher, Janka Flaska, Homaira Hanif, Sarah Henley-Shepard, Marissa Hildebrandt, Esther Johnston, Gifty Kwakye, Lindsay Mathieson, Siri Michel, Karen Milch, Sarah Murray, Catherine Packer, Evan Russell, Elena Semenova, Fatima Sharif, and Michelle Vanstone for their involvement in the systematic literature review and historical event review compilation. We would also like to thank John McGready for biostatistical support, Claire Twose assistance in designing and implementing the systematic literature review, and Hannah Tappis and Bhakti Hansoti for their support in the revision process.

REFERENCES

1. Jonkman SN & Kelman I. An analysis of the causes and circumstances of flood disaster deaths. Disasters. 2005; 29(1):75-97

2. Noji E. Public Health Issues in Disasters. Critical Care Medicine. 2000; 33(1):S29-S33

3. Jonkman, SN. Global perspectives on loss of human life caused by floods. Natural Hazards. 2005; 34(2): 151-175.

4. EM-DAT the International Disaster Database. (2009). Retrieved Jan 12, 2009, from www.emdat.be/

5. Hunt, JC. Inland and coastal flooding: Developments in prediction and prevention. Philosophical transactions. Series A, Mathematical, Physical, and Engineering Sciences. 2005; 363(1831): 1475-1491.

6. Tobin GA and Montz, BE (Eds.). Natural hazards: Explanations and integration. New York: Guilford Press, 1997.

7. Ashley ST and Ashley WA . Flood Fatalities in the United States. Journal of Applied Meteorology and Climatology. 2008; 47:805-818

8. Beinin L. Medical Consequences of Natural Disasters. Berlin: Springer-Verlag. 1985.

9. Alajo SO, Nakavuma J, Erume J. Cholera in endemic districts of Uganda during El Nino rains: 2002-2003. African Health Sciences. 2006; 6(2):93- 97.

10. Li, SQ, Tan HZ, Li, XL, et al. A study on the health status of residents affected by flood disaster. Chinese Journal of Epidemiology. 2004; 25:36-39.

11. French JG, Ing R, Von Allmen S, et al. Mortality from flash flood: A review of National Weather Service Reports. Public Health Reports. 1983; 98:584-588

12. Kahn M. The death toll from natural disasters: the role of income, geography, and institutions. The Review of Economics and Statistics. 2005; 87(2):271-284.

13. Dartmouth Flood Observatory: Global Active Archive of Large Flood Events. Available at http:///www.dartmouth.edu/~floods/Archives/index.html. Accessed July 2007 and August 2009.

14. StataCorp. Chicago, IL. Stata Statistical Software: Release 11.0.

15. Duclos P, Vidonne O, Beuf P, et al. (1991). Flash flood disaster--Nimes, France, 1988. European Journal of Epidemiology, 7(4), 365-371.

16. Siddique AK. Baqui AH. Eusof A, et al. 1988 floods in Bangladesh - pattern of illness and causes of death. Journal of Diarrhoeal Diseases Research. 1991; 9(4):310-314.

17. Centers for Disease Control and Prevention (CDC). Morbidity surveillance following the Midwest flood--Missouri, 1993. MMWR. Morbidity and Mortality Weekly Report. 1993; 42(41):797-798.

18. Duke C, Bon E, Reeves J, et al. (1994). Flood-related mortality - Georgia, July 4-14, 1994. MMWR. 1994; 43(29):526-530

19. Staes C, Orengo JC, Malilay J, et al. Deaths due to flash floods in Puerto Rico, January 1992: Implications for prevention. International Journal of Epidemiology. 1994; 23(5): 968-975.

20. Centers for Disease Control and Prevention (CDC). Storm-related mortality--central Texas, October 17-31, 1998. MMWR. Morbidity and Mortality Weekly Report. 2000; 49(7): 133-135.

21. Rashid SF. The urban poor in Dhaka city: Their struggles and coping strategies during the floods of 1998. Disasters. 2000; 24(3); 240-253.

22. Yale JD. Cole TB, Garrison HG, et al. Motor vehicle-related drowning deaths associated with inland flooding after Hurricane Floyd: A field investigation. Traffic Injury Prevention. 2003; 4(4):279-284.

23. Pradhan EK, West KP, Katz J, et al. Risk of flood-related mortality in Nepal. Disasters. 2007; 31(1):57-70.

24. Baxter PJ. The east coast big flood, 31 January-1 February 1953: A summary of the human disaster. Philosophical Transactions: Mathematical, Physical and Engineering Sciences (Series A). 2005; 363(1831): 1293-1312.

25. Gerritsen H. What happened in 1953? The big flood in the Netherlands in retrospect. Philosophical Transactions: Mathematical, Physical and Engineering Sciences (Series A), 2005; 363(1831):1271-1291.

26. Spencer J and Myer R. A population and economic overview of Cambria County, Pennsylvania following the 1977 Johnstown flood. Disaster Prevention and Management. 2007; 16(2):259-264.

27. Grigg NS, Doesken NJ, Frick DM, et al. Fort Collins flood 1997: Comprehensive view of an extreme event. Journal of Water Resources Planning and Management. 1999; 125(5):255-262.

28. Jonkman SN, Maaskant B, Boyd E, et al. Loss of life caused by the flooding of New Orleans after Hurrican Katrina: Analysis of the relationship between flood characteristics and mortality. Risk Analysis. 2009; 29(5):676-698.

29. Ahern M, Kovats RS, Wilkinson P, et al. Global health impact of floods: epidemiological evidence. Epidemiologic Reviews; 2005 27(1):36-46.

30. Beyhun N, Altintas KH and Noji, E. Analysis of registered floods in Turkey. International Journal of Disaster Medicine. 2005; 3(1-4): 50-54.

31. Tarhule A. Damaging rainfall and flooding: The other Sahel hazards. Climatic Change. 2005; 72(3), 355-377.

32. Ogden CL, Gibbs-Scharf LI, Kohn MA et al. (2001). Emergency health surveillance after severe flooding in Louisiana, Prehospital and Disaster Medicine. 1995; 16(3):138-144.

33. Centers for Disease Control and Prevention (CDC). Morbidity surveillance following the Midwest flood--Missouri, 1993. MMWR. 1993; 42(41):97-798.

34. Cariappa MP and Khanduri P. Health emergencies in large populations: The Orissa experience. Medical Journal of the Armed Forces of India. 2003; 59(4):286-289.

35. Schnitzler J, Benzler J, Altmann D, et al. Survey on the population's needs and the public health response during floods in Germany 2002. Public Health Management and Practice. 2007; 13(5):461-464.

36. Biswas A, Rahman A, Mashreky S, et al. Unintentional injuries and parental violence against children during flood: a study in rural Bangladesh. Rural and Remote Health. 2010; 10:1199.

37. Sanchez-Crispin A an d Propin-Frejomil E. Social and economic dimensions of the 1998 extreme floods in coastal Chiapas, Mexico. IAHS-AISH Publication. 2002;(271):385-390.

38. Berz, G. Flood disasters: Lessons from the past - worries for the future. Proceedings of the Institution of Civil Engineers: Water, Maritime and Energy. 2000;142(1):3-8.

39. Graham W. A procedure for estimating loss of life caused by dam failure. Dam Safety Office Report 99-06. USBR: Denver CO, 1999.

40. Stromberg, D. Natural Disasters, Economic Development, and Humanitarian Aid. The Journal of Economic Perspectives. 2007; 21(3):199-222.

41. UNISDR. Linking disaster risk reduction and poverty reduction: good practices and lessons learned. Geneva: UNISDR, 2008.

42. Haque C. Perspectives of natural disasters in East and South Asia and the Pacific Island States: socioeconomic correlates and needs assessment. Natural Hazards. 2003; 29:465-483.

43. Lllewellyn M. Floods and Tsunamis. Surgical clinics of North America. 2006; 86:557-578.

44. French J, Ing R, Von Allmen S, et al. Mortality from flash floods: A review of national weather service reports, 1969-81. Public Health Reports. 1983; 98(6):584-588

45. Mooney LE. Applications and implications of fatality statistics to the flash flood problem. The American Meteorological Society Fifth Conference on Hydrometeorology, 1983. Preprints, p127–129.

46. Coates L. Flood fatalities in Australia, 1788–1996. Australian Geographer. 1999; 30:391–408.

47. Chowdhury AR, Mushtaque AU, Bhuyia AY, et al. The Bangladesh Cyclone of 1991: Why So Many People Died. Disasters. 1993; 17(4):291–304.

48. Jonkman SN and Vrijling JK. Loss of life due to floods. Journal of Flood Risk Management. 2008; 1: 43-56.

49. FitzGerald G, Du W, Jamal A, et al. Flood fatalities in contemporary Australia (1997-2008). Emergency Medicine Australasia. 2010; 22:80-186.

50. Guha-Sapir, D. Natural and man-made disasters: the vulnerability of women-headed households and children without families. World Health Statistics Quarterly. 1993; 46:227-233.

51. Creel L. Ripple effects: Population and coastal regions. Population Reference Bureau. 2003.URL:http://www.prb.org/Publications/PolicyBriefs/RippleEffectsPopulationandCoastalRegions.aspx.

52. UN Habitat. State of the World's Cities 2010/2011: Bridging the Urban Divide. March 18, 2010. UN Human Settlements Programme: Nairobi, Kenya.

53.   Janerich D T, StarkS AD, Greenwald P, et al. Increased leukemia, lymphoma, and spontaneous abortion in Western New York following a flood disaster. Public Health Reports. 1981; 96(4):350-356.

54.   Bich TH, Quang LN, Ha LTT, Hanh TTD, Guha-Sapir D. Impacts of flood on health: epidemiologic evidence from Hanoi, Vietnam. Global Healh Action. 2011; 4:6356.

55.   Mortality from tornadoes, hurricanes, and floods. Statistical Bulletin of the Metropolitan Life Insurance Company. 1974; 55:4-7.

56.   Avakyan AB and Istomina MN. Floods in the world late in the XX century. Water Resources. 2000; 27(5): 469-475.

57.   Guzzetti F, Stark CP, Salvati P. Evaluation of flood and landslide risk to the population of Italy. Environmental Management. 2005; 36(1):15-36.

58.   Lastoria B, Simonetti MR, Casaioli M, et al. Socio-economic impacts of major floods in Italy from 1951 to 2003. Advances in Geosciences. 2006; 7:223-229.

59.   United Nations Development Program. Human Development Reports. (2009). Available at: http://hdrstats.undp.org/en/indicators/#G. Accessed May 18, 2010.

Chapter 6

Epidemics After Natural Disasters

John T. Watson[1,*], Michelle Gayer,[*] and Maire A. Connolly[*]

[*]World Health Organization, Geneva, Switzerland

[1]Disease Control in Humanitarian Emergencies, Communicable Diseases Cluster, World Health Organization, 1211 Geneva, Switzerland

ABSTRACT

The relationship between natural disasters and communicable diseases is frequently misconstrued. The risk for outbreaks is often presumed to be very high in the chaos that follows natural disasters, a fear likely derived from a perceived association between dead bodies and epidemics.

However, the risk factors for outbreaks after disasters are associated primarily with population displacement. The availability of safe water and sanitation facilities, the degree of crowding, the underlying health status of the population, and the availability of healthcare services all interact within the context of the local disease ecology to influence the risk for communicable diseases and death in the affected population. We outline the risk factors for outbreaks after a disaster, review the communicable diseases likely to be important, and establish priorities to address communicable diseases in disaster settings.

Natural disasters are catastrophic events with atmospheric, geologic, and hydrologic origins. Disasters include earthquakes, volcanic eruptions, landslides, tsunamis, floods, and drought. Natural disasters can have rapid or slow onset, with serious health, social, and economic consequences. During the past 2 decades, natural disasters have killed millions of people, adversely affected the lives of at least 1 billion more people, and resulted in substantial economic damages (1). Developing countries are disproportionately affected because they lack resources, infrastructure, and disaster-preparedness systems.

Deaths associated with natural disasters, particularly rapid-onset disasters, are overwhelmingly due to blunt trauma, crush-related injuries, or drowning. Deaths from communicable diseases after natural disasters are less common.

DEAD BODIES AND DISEASE

The sudden presence of large numbers of dead bodies in the disaster-affected area may heighten concerns of disease outbreaks (2), despite the absence of evidence that dead bodies pose a risk for epidemics after natural disasters (3). When death is directly due to the natural disaster, human remains do not pose a risk for outbreaks (4). Dead bodies only pose health risks in a few situations that require specific precautions, such as deaths from cholera (5) or hemorrhagic fevers (6). Recommendations for management of dead bodies are summarized in the Table.

Table 1: Principles for management of dead bodies*

Mass management of dead bodies is often based on the false belief that they represent an epidemic hazard if not buried or burned immediately.
Burial is preferable to cremation in mass casualty situations.
Every effort should be made to identify the bodies. Mass burial should be avoided if at all possible.
Families should have the opportunity (and access to materials) to conduct culturally appropriate funerals and burials according to social custom.
Where existing facilities such as graveyards or crematoria are inadequate, alternative locations or facilities should be provided.
For workers routinely handling bodies, ensure
Universal precautions for blood and body fluids
Use and correct disposal of gloves
Use of body bags if available
Hand-washing with soap after handling bodies and before eating
Disinfection of vehicles and equipment
Bodies do not need disinfection before disposal (except in cases of cholera, shigellosis, or hemorrhagic fever)
Bottom of any grave is >1.5 m above the water table, with a 0.7-m unsaturated zone

*Adapted from Morgan (3).

Despite these facts, the risk for outbreaks after disasters is frequently exaggerated by both health officials and the media. Imminent threats of epidemics remain a recurring theme of media reports from areas recently affected by disasters, regardless of attempts to dispel these myths (2,3,7).

DISPLACEMENT: PRIMARY CONCERN

The risk for communicable disease transmission after disasters is associated primarily with the size and characteristics of the population displaced, specifically the proximity of safe water and functioning latrines, the nutritional status of the displaced population, the level of immunity to vaccine-preventable diseases such as measles, and the access to healthcare services (8). Outbreaks are less frequently reported in disaster-affected populations than in conflict-affected populations, where two thirds of deaths may be from communicable diseases (9). Malnutrition increases the risk for death from communicable diseases and is more common in conflict-affected populations, particularly if their displacement is related to long-term conflict (10).

Although outbreaks after flooding (11) have been better documented than those after earthquakes, volcanic eruptions, or tsunamis (12), natural disasters (regardless of type) that do not result in population displacement are rarely associated with outbreaks (8). Historically, the large-scale displacement of populations as a result of natural disasters is not common (8), which likely contributes to the low risk for outbreaks overall and to the variability in risk among disasters of different types.

RISK FACTORS FOR COMMUNICABLE DISEASE TRANSMISSION

Responding effectively to the needs of the disaster-affected population requires an accurate communicable disease risk assessment. The efficient use of humanitarian funds depends on implementing priority interventions on the basis of this risk assessment.

A systematic and comprehensive evaluation should identify 1) endemic and epidemic diseases that are common in the affected area; 2) living conditions of the affected population, including number, size, location, and density of settlements; 3) availability of safe water and adequate sanitation facilities; 4) underlying nutritional status and immunization coverage among the population; and 5) degree of access to healthcare and to effective case management.

COMMUNICABLE DISEASES ASSOCIATED WITH NATURAL DISASTERS

The following types of communicable diseases have been associated with populations displaced by natural disasters. These diseases should be considered when postdisaster risk assessments are performed.

Water-related Communicable Diseases

Access to safe water can be jeopardized by a natural disaster. Diarrheal disease outbreaks can occur after drinking water has been contaminated and have been reported after flooding and related displacement. An outbreak of diarrheal disease after flooding in Bangladesh in 2004 involved >17,000 cases; *Vibrio cholerae* (O1 Ogawa and O1 Inaba) and enterotoxigenic *Escherichia coli* were isolated (*13*). A large (>16,000 cases) cholera epidemic (O1 Ogawa) in West Bengal in 1998 was attributed to preceding floods (*14*), and floods in Mozambique in January–March 2000 led to an increase in the incidence of diarrhea (*15*).

In a large study undertaken in Indonesia in 1992–1993, flooding was identified as a significant risk factor for diarrheal illnesses caused by *Salmonella enterica* serotype Paratyphi A (paratyphoid fever) (*16*). In a separate evaluation of risk factors for infection with *Cryptosporidium parvum* in Indonesia in 2001–2003, case-patients were >4× more likely than controls to have been exposed to flooding (*17*).

The risk for diarrheal disease outbreaks following natural disasters is higher in developing countries than in industrialized countries (*8,11*). In Aceh Province, Indonesia, a rapid health assessment in the town of Calang 2 weeks after the December 2004 tsunami found that 100% of the survivors drank from unprotected wells and that 85% of residents reported diarrhea in the previous 2 weeks (*18*). In Muzaffarabad, Pakistan, an outbreak of acute watery diarrhea occurred in an unplanned, poorly equipped camp of 1,800 persons after the 2005 earthquake. The outbreak involved >750 cases, mostly in adults, and was controlled after adequate water and sanitation facilities were provided (*19*). In the United States, diarrheal illness was noted after Hurricanes Allison (*20*) and Katrina (*21–23*), and norovirus, *Salmonella*, and toxigenic and no toxigenic *V. cholerae* were confirmed among Katrina evacuees.

Hepatitis A and E are also transmitted by the fecal-oral route, in association with lack of access to safe water and sanitation. Hepatitis A is endemic in most developing countries, and most children are exposed and develop immunity at an early age. As a result, the risk for large outbreaks is usually low in these settings. In hepatitis E–endemic areas, outbreaks frequently follow heavy rains and floods; the illness is generally mild and self-limited, but in pregnant women case-fatality rates can reach 25% (*24*). After the 2005 earthquake in Pakistan, sporadic hepatitis E cases and clusters were common in areas with poor access to safe water. Over 1,200 cases of acute jaundice, many confirmed as hepatitis E, occurred among the displaced (*25*). Clusters of both hepatitis A and hepatitis E were noted in Aceh after the December 2004 tsunami (*26*).

Leptospirosis is an epidemic-prone zoonotic bacterial disease that can be transmitted by direct contact with contaminated water. Rodents shed large amounts of leptospires in their urine, and transmission occurs through contact of the skin and mucous membranes with water, damp soil or vegetation (such as sugar cane), or mud contaminated with rodent urine. Flooding facilitates spread of the organism because of the proliferation of rodents and the proximity of rodents to humans on shared high ground. Outbreaks of leptospirosis occurred in Taiwan, Republic of China, associated with Typhoon Nali in 2001 (*27*); in Mumbai, India, after flooding in 2000 (*28*); in Argentina after flooding in 1998 (*29*); and in the Krasnodar region of the Russian Federation in 1997 (*30*). After a flooding-related outbreak of leptospirosis in Brazil in 1996, spatial analysis indicated that incidence rates of leptospirosis doubled inside the flood-prone areas of Rio de Janeiro (*31*).

Diseases Associated with Crowding

Crowding is common in populations displaced by natural disasters and can facilitate the transmission of communicable diseases. Measles and the risk for transmission after a natural disaster are dependent on baseline immunization coverage among the affected population, and in particular among children <15 years of age. Crowded living conditions facilitate measles transmission and necessitate even higher immunization coverage levels to prevent outbreaks (32). A measles outbreak in the Philippines in 1991 among persons displaced by the eruption of Mt. Pinatubo involved >18,000 cases (33). After the tsunami in Aceh, a cluster of measles involving 35 cases occurred in Aceh Utara district, and continuing sporadic cases and clusters were common despite mass vaccination campaigns (26). In Pakistan, after the 2005 South Asia earthquake, sporadic cases and clusters of measles (>400 clinical cases in the 6 months after the earthquake) also occurred (25).

Neisseria meningitidis meningitis is transmitted from person to person, particularly in situations of crowding. Cases and deaths from meningitis among those displaced in Aceh and Pakistan have been documented (25,26). Prompt response with antimicrobial prophylaxis, as occurred in Aceh and Pakistan, can interrupt transmission. Large outbreaks have not been recently reported in disaster-affected populations but are well-documented in populations displaced by conflict (34).

Acute respiratory infections (ARI) are a major cause of illness and death among displaced populations, particularly in children <5 years of age. Lack of access to health services and to antimicrobial agents for treatment further increases the risk for death from ARI. Risk factors among displaced persons include crowding, exposure to indoor cooking using open flame, and poor nutrition. The reported incidence of ARI increased 4-fold in Nicaragua in the 30 days after Hurricane Mitch in 1998 (35), and ARI accounted for the highest number of cases and deaths among those displaced by the tsunami in Aceh in 2004 (26) and by the 2005 earthquake in Pakistan (25).

Vectorborne Diseases

Natural disasters, particularly meteorologic events such as cyclones, hurricanes, and flooding, can affect vector-breeding sites and vectorborne disease transmission. While initial flooding may wash away existing mosquito-breeding sites, standing water caused by heavy rainfall or overflow of rivers can create new breeding sites. This situation can result (with typically some weeks' delay) in an increase of the vector population and potential for disease transmission, depending on the local mosquito vector species and its preferred habitat. The crowding of infected and susceptible hosts, a weakened public health infrastructure, and interruptions of ongoing control programs are all risk factors for vectorborne disease transmission (36).

Malaria outbreaks in the wake of flooding are a well-known phenomenon. An earthquake in Costa Rica's Atlantic Region in 1991 was associated with changes in habitat that were beneficial for breeding and preceded an extreme rise in malaria cases (37). Additionally, periodic flooding linked to El Niño–Southern Oscillation has been associated with malaria epidemics in the dry coastal region of northern Peru (38).

Dengue transmission is influenced by meteorologic conditions, including rainfall and humidity, and often exhibits strong seasonality. However, transmission is not directly associated with flooding. Such events may coincide with periods of high risk for transmission and may be exacerbated by increased availability of the vector's breeding sites (mostly artificial containers) caused by disruption of basic water supply and solid waste disposal services. The risk for outbreaks can be influenced by other complicating factors, such as changes in human behavior (increased exposure to mosquitoes while sleeping outside, movement from dengue-nonendemic to -endemic areas, a pause in disease control activities, overcrowding) or changes in the habitat that promote mosquito breeding (landslide, deforestation, river damming, and rerouting of water).

Other Diseases Associated with Natural Disasters

Tetanus is not transmitted person to person but is caused by a toxin released by the anaerobic tetanus bacillus *Clostridium tetani*. Contaminated wounds, particularly in populations where vaccination coverage levels are low, are associated with illness and death from tetanus. A cluster of 106 cases of tetanus, including 20 deaths, occurred in Aceh and peaked 2-1/2 weeks after the tsunami (26). Cases were also reported in Pakistan following the 2005 earthquake (25).

An unusual outbreak of coccidiomycosis occurred after the January 1994 Southern California earthquake. The infection is not transmitted person to person and is caused by the fungus *Coccidioides immitis*, which is found in soil in certain semiarid areas of North and South America. This outbreak was associated with exposure to increased levels of airborne dust subsequent to landslides in the aftermath of the earthquake (*39*).

Disaster-Related Interruption of Services

Power cuts related to disasters may disrupt water treatment and supply plants, thereby increasing the risk for waterborne diseases. Lack of power may also affect proper functioning of health facilities, including preservation of the vaccine cold chain. An increase in diarrheal illness in New York City followed a massive power outage in 2003. The blackout left 9 million people in the area without power for several hours to 2 days. Diarrhea cases were widely dispersed and detected by using nontraditional surveillance techniques. A case-control study performed as part of the outbreak investigation linked diarrheal illness with the consumption of meat and seafood after the onset of the power outage, when refrigeration facilities were widely interrupted (*40*).

DISCUSSION

Historically, fears of major disease outbreaks in the aftermath of natural disasters have shaped the perceptions of the public and policymakers. These expectations, misinformed by associations of disease with dead bodies, can create fear and panic in the affected population and lead to confusion in the media and elsewhere.

The risk for outbreaks after natural disasters is low, particularly when the disaster does not result in substantial population displacement. Communicable diseases are common in displaced populations that have poor access to basic needs such as safe water and sanitation, adequate shelter, and primary healthcare services. These conditions, many favorable for disease transmission, must be addressed immediately with the rapid reinstatement of basic services. Assuring access to safe water and primary healthcare services is crucial, as are surveillance and early warning to detect epidemic-prone diseases known to occur in the disaster-affected area. A comprehensive communicable disease risk assessment can determine priority diseases for inclusion in the surveillance system and prioritize the need for immunization and vector-control campaigns. Five basic steps that can reduce the risk for communicable disease transmission in populations affected by natural disasters are summarized in an (Appendix Table).

Disaster-related deaths are overwhelmingly caused by the initial traumatic impact of the event. Disaster-preparedness plans, appropriately focused on trauma and mass casualty management, should also take into account the health needs of the surviving disaster-affected populations. The health effects associated with the sudden crowding of large numbers of survivors, often with inadequate access to safe water and sanitation facilities, will require planning for both therapeutic and preventive interventions, such as the rapid delivery of safe water and the provision of rehydration materials, antimicrobial agents, and measles vaccination materials.

Surveillance in areas affected by disasters is fundamental to understanding the impact of natural disasters on communicable disease illness and death. Obtaining relevant surveillance information in these contexts, however, is frequently challenging. The destruction of the preexisting public health infrastructure can aggravate (or eliminate) what may have been weak predisaster systems of surveillance and response. Surveillance officers and public health workers may be killed or missing, as in Aceh in 2004. Population displacement can distort census information, which makes the calculation of rates for comparison difficult. Healthcare during the emergency phase is often delivered by a wide range of national and international actors, which creates coordination challenges. Also, a lack of predisaster baseline surveillance information can lead to difficulties in accurately differentiating epidemic from background endemic disease transmission.

Although postdisaster surveillance systems are designed to rapidly detect cases of epidemic-prone diseases, interpreting this information can be hampered by the absence of baseline surveillance data and accurate denominator values. Detecting cases of diseases that occur endemically may be interpreted (because of absence of background data) as an early epidemic. The priority in these settings, however, is rapid implementation of control measures when cases of epidemic-prone diseases are detected. Despite these challenges, continued detection of and response to communicable diseases are essential to monitor the incidence of diseases, to document their effect, to respond with control measures when needed, and to better quantify the risk for outbreaks after disasters.

ACKNOWLEDGMENTS

We thank Pamela Mbabazi, Jorge Castilla, Andre Griekspoor, José Hueb, Dominique Legros, David Meddings, Mike Nathan, Aafje Rietveld, and Peter Strebel for their support and assistance with the preparation of this manuscript.

REFERENCES

1. United Nations Cultural Scientific and Cultural Organization [homepage on the internet]. Paris. About natural disasters. [cited 2006 Aug 10]. Available fromhttp://www.unesco.org/science/disaster/about_disaster.shtml

2. de Ville de Goyet C. Epidemics caused by dead bodies: a disaster myth that does not want to die. Rev Panam Salud Publica 2004;15:297–9

3. Morgan O Infectious disease risks from dead bodies following natural disasters. Rev Panam Salud Publica 2004;15:307–11 10.1590/S1020-49892004000500004

4. Management of dead bodies in disaster situations. (PAHO disaster manuals and guidelines on disaster series, no. 5).Washington: Pan American Health Organization; 2004

5. Sack RB, Siddique AK Corpses and the spread of cholera. Lancet 1998;352:1570 10.1016/S0140-6736(05)61040-9

6. Boumandouki P, Formenty P, Epelboin A, Campbell P, Atsangandoko C, Allarangar Y, et al. Clinical management of patients and deceased during the Ebola outbreak from October to December 2003 in Republic of Congo [article in French] Bull Soc Pathol Exot2005;98:218–23

7. de Ville de Goyet C. Stop propagating disaster myths. Lancet 2000;356:762–4 10.1016/S0140-6736(00)02642-8

8. Noji E, ed. Public health consequences of disasters. New York: Oxford University Press; 1997

9.   Noji EK Public health in the aftermath of disasters. BMJ 2005;330:1379–81 10.1136/bmj.330.7504.1379

10.   Spiegel PB Differences in world responses to natural disasters and complex emergencies. JAMA 2005;293:1915–8 10.1001/jama.293.15.1915

11.   Ahern M, Kovats RS, Wilkinson P, Few R, Matthies F Global health impacts of floods: epidemiologic evidence. Epidemiol Rev 2005;27:36–46 10.1093/epirev/mxi004

12.   Floret N, Viel J-F, Mauny F, Hoen B, Piarroux R Negligible risk for epidemics after geophysical disasters. [PMID: 16704799] Emerg Infect Dis 2006;12:543–8

13.   Qadri F, Khan AI, Faruque ASG, Begum YA, Chowdhury F, Nair GB, et al.Enterotoxigenic *Escherichia coli* and *Vibrio cholerae* diarrhea, Bangladesh. Emerg Infect Dis 2005;11:1104–7

14.   Sur D Severe cholera outbreak following floods in a northern district of West Bengal.Indian J Med Res 2000;112:178–82

15.   Kondo H, Seo N, Yasuda T, Hasizume M, Koido Y, Ninomiya N, et al. Post-flood–infectious diseases in Mozambique. Prehosp Disaster Med 2002;17:126–33

16.   Vollaard AM, Ali S, van Asten HA, Widjaja S, Visser LG, Surjadi C, et al. Risk factors for typhoid and paratyphoid fever in Jakarta, Indonesia. JAMA 2004;291:2607–15 10.1001/jama.291.21.2607

17.   Katsumata T, Hosea D, Wasito EB, Kohno S, Hara K, Soeparto P, et al.Cryptosporidiosis in Indonesia: a hospital-based study and a community-based survey.Am J Trop Med Hyg 1998;59:628–32

18.   Brennan RJ, Kimba K Rapid health assessment in Aceh Jaya District, Indonesia, following the December 26 tsunami. Emerg Med Australas 2005;17:341–50 10.1111/j.1742-6723.2005.00755.x

19.   World Health Organization Acute water diarrhea outbreak. Weekly Morbidity and Mortality Report. 2005;1:6. [cited 2006 Aug 10].Available fromhttp://www.who.int/hac/crises/international/pakistan_earthquake/sitrep/FINAL_WMMR_Pakistan_1_December_06122005.pdf

20.   Waring SC, Reynolds KM, D'Souza G, Arafat RR Rapid assessment of household needs in the Houston area after Tropical Storm Allison. Disaster Manag Response 2002; (Sep):3–9

21. Centers for Disease Control and Prevention (CDC) Norovirus outbreak among evacuees from hurricane Katrina–Houston, Texas, September 2005. MMWR Morb Mortal Wkly Rep 2005;54:1016–8

22. Centers for Disease Control and Prevention Infectious disease and dermatologic conditions in evacuees and rescue workers after Hurricane Katrina—multiple states, August–September, 2005. MMWR Morb Mortal Wkly Rep 2005;54:961–4

23. Centers for Disease Control and Prevention Two cases of toxigenic Vibrio cholerae O1 infection after Hurricanes Katrina and Rita—Louisiana, October 2005. MMWR Morb Mortal Wkly Rep 2006;55:31–2

24. Aggarwal R, Krawczynski K Hepatitis E: an overview and recent advances in clinical and laboratory research. J Gastroenterol Hepatol 2000;15:9–20 10.1046/j.1440-1746.2000.02006.x

25. World Health Organization Acute jaundice syndrome. Weekly Morbidity and Mortality Report. 2006;23:8. [cited 2006 Aug 10].Available fromhttp://www.who.int/hac/crises/international/pakistan_earthquake/sitrep/Pakistan_WMMR_VOL23_03052006.pdf

26. World Health Organization Epidemic-prone disease surveillance and response after the tsunami in Aceh Province, Indonesia. Wkly Epidemiol Rec 2005;80:160–4

27. Yang HY, Hsu PY, Pan MJ, Wu MS, Lee CH, Yu CC, et al. Clinical distinction and evaluation of leptospirosis in Taiwan—a case-control study. J Nephrol 2005;18:45–53

28. Karande S, Bhatt M, Kelkar A, Kulkarni M, De A, Varaiya A An observational study to detect leptospirosis in Mumbai, India, 2000. Arch Dis Child 2003;88:1070–5 10.1136/adc.88.12.1070

29. Vanasco NB, Fusco S, Zanuttini JC, Manattini S, Dalla Fontana ML, Prez J, et al.Outbreak of human leptospirosis after a flood in Reconquista, Santa Fe, 1998 [article in Spanish] Rev Argent Microbiol 2002;34:124–31

30. Kalashnikov IA, Mezentsev VM, Mkrtchan MO, Grizhebovskii GM, Briukhanova GD Features of leptospirosis in the Krasnodar Territory [article in Russian] Zh Mikrobiol Epidemiol Immunobiol 2003; (Nov-Dec):68–71

31. Barcellos C, Sabroza PC The place behind the case: leptospirosis risks and associated environmental conditions in a flood-related outbreak in Rio de Janeiro. Cad Saude Publica2001;17(Suppl):59–67

32. Marin M, Nguyen HQ, Langidrik JR, Edwards R, Briand K, Papania MJ, et al.Measles transmission and vaccine effectiveness during a large outbreak on a densely populated island: implications for vaccination policy. Clin Infect Dis 2006;42:315–9 10.1086/498902

33. Surmieda MR, Lopez JM, Abad-Viola G, Miranda ME, Abellanosa IP, Sadang RA, et al. Surveillance in evacuation camps after the eruption of Mt. Pinatubo, Philippines.MMWR CDC Surveill Summ 1992;41:963

34. Gaspar M, Leite F, Brumana L, Felix B, Stella AA Epidemiology of meningococcal meningitis in Angola, 1994–2000. Epidemiol Infect 2001;127:421–4 10.1017/S0950268801006318

35. Campanella N Infectious diseases and natural disasters: the effects of Hurricane Mitch over Villanueva municipal area, Nicaragua. Public Health Rev 1999;27:311–9

36. Lifson AR Mosquitoes, models, and dengue. Lancet 1996;347:1201–2 10.1016/S0140-6736(96)90730-8

37. Saenz R, Bissell RA, Paniagua F Post-disaster malaria in Costa Rica. Prehosp Disaster Med 1995;10:154–60

38. Gagnon AS, Smoyer-Tomic KE, Bush AB The El Nino southern oscillation and malaria epidemics in South America. Int J Biometeorol 2002;46:81–9 10.1007/s00484-001-0119-6

39. Schneider E, Hajjeh RA, Spiegel RA, Jibson RW, Harp EL, Marshall GA, et al. A coccidiomycosis outbreak following the Northridge, Calif, earthquake. JAMA1997;277:904–8 10.1001/jama.277.11.904

40. Marx MA, Rodriguez CV, Greenko J, Das D, Heffernan R, Karpati AM, et al.Diarrheal illness detected through syndromic surveillance after a massive power outage: New York City, August 2003. Am J Public Health 2006;96:547–53 10.2105/AJPH.2004.061358

Chapter 7

Flash Flooding in Attika, Greece: Climatic Change or Urbanization

Ourania Lasda,[1] Angela Dikou,[1,2] and Evangelos Papapanagiotou[1]

[1]Department of Environment, University of the Aegean, Mytilene, 81100 Greece

[2]Department of Ichthyology and Aquatic Sciences, University of Thessaly, Volos, 38446 Greece, Ourania Lasda.

FLOODING RECORDS IN ATTIKA

Flooding constituted the second most frequent natural disaster in Greece during 1928–2005 (15 episodes; 23.4% of total) after earthquakes; it led to 78 deaths, 10,990 affected people and 719,518,000 US$ damage and repair costs (World Health Organisation 2005). Nevertheless, the

28 episodes of flooding in Attika Prefecture (Fig. 1) cost more human lives (182 people) during the last century (1887–2005) than earthquakes (18 people) while the cost in human lives due to flooding for the whole country during the same period was 220 people (Nicolaidou and Hadjichristou 1995; Lasda 2005.[1]

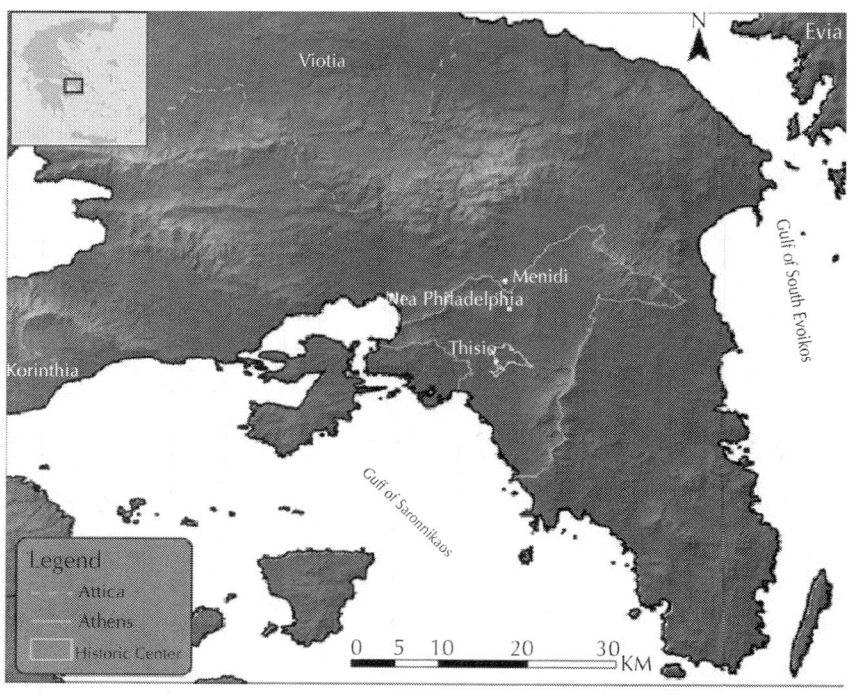

Figure 1: Map of Attika Prefecture in Greece.

LAND USE AND HYDROGRAPHIC CHANGES

wo events exerted significant influence on Attika's population during the twentieth century: the inor Asia emigration in 1922, which led to the entry of thousands of refugees to Attika, and the internal migration to the capital of Greece that followed the end of World War II. The Attika basin gathered only an 18% (1,394,922 people) of Greece's population in 1951, which reached a 35% (3,062,278 people) in 1981. The population of Attika basin tripled during 1940–2001 (188.8% increase, 3,292,189 people) (Greek National Statistics Service 2001).

The influx of people at the 3,808 km^2 Attika basin boosted construction activities. Thus, Attika's urbanized area increased from 17.7% in 1945 to 39.6% in 1973 and to 68.5% in 1995 (Sioras 2003). The increase in urbanized area led to concomitant decrease in cumulative cultivated, forested and shrub areas (cumulative percent area cover was 81.3% in 1945; 42.5% in 1973; 31.0% in 1995). The development paradigm followed for several decades lacked organization and wisdom leading to congestion and overuse of natural resources in urban areas of Attika. Human interference at suburban areas of Attika included land clearance, agricultural abandonment, forest fires, unplanned expansion of urban areas, and rubble-filing for the creation of roads and plots.

Dramatic were also the gradual changes in the hydrographic network of Attika, which led to the reduction of the total length of its active branches. This reduction had started already since 1893–1945 (from 1.277 to 858 km, 33.6% reduction), continued during 1945–1973 (from 858 to 734 km, 14.5% reduction) and culminated during 1973–1995 (from 734 to 434 km, 40.9% reduction) (Sioras 2003). Thus, there was a 66.4% reduction in the hydrographic network during 1893–1995 while the ratio of active to inactive streams was 0.5 in 1995. The evolution of the state of the hydrographic network may be pictured as small and large streams being covered or modified, totally or partially with construction works of various types and scales within their watercourse (but mostly at their banks) and being inflicted upon with trespasses and unauthorized construction (Bitsika 2003). Some streams were modified or blocked in order to avoid unpleasant consequences, such as flash floods or stench. Construction on top of, modification, and blockage of streams were identified, however, as modernization and a trend of discarding what was related to the past, the under-development and the low standards of living. Furthermore, solid waste discard in streams became a usual phenomenon leading to the creation of pollution foci, which are hazardous to public health, and reduction of the channeling of active streams (Sioras 2003).

The aforementioned changes in land use and hydrographic network led to reduction in the infiltration of water and the enrichment capacity of the aquatic resources and to increase in erosion, surface runoff and locality-specific vulnerability to flash flooding depending on distance from streams and slope of land at Attika basin (Sioras 2003; sakiris 2004; imikou 2005; Ministry of Planning, Public Works and Environment 2005).

CLIMATIC CHANGE

Global climate change induced by increased greenhouse gas concentrations has been widely recognized and is often cited as involving changes in temperature and precipitation patterns. Precipitation exhibited significant declining trends in higher latitude regions in the Northern Hemisphere given precipitation records of the last century (Dore 2005). While countries of northern Europe have experienced increase in precipitation during the last decades, countries of south Europe bordering the Mediterranean Sea (including Greece) have experienced reduction in precipitation (Tsiourtis 2002). Time series analysis of precipitation in Athens during 1925–1999 revealed a decreasing (by 10%), yet, not significant trend (Feidas and Lalas 2000).

Nevertheless, the flash flooding episodes of the 8th July, 3rd August and 18th August 2002 in Attika have been attributed to extreme, for the summer season, rainfall intensities (Bitsika and Tratsa 2002). The rainfall monitoring station at Thisio recorded 91, 56 and 48 mm of rainfall during 24 h, respectively, during the aforementioned flash flooding episodes while the rainfall monitoring station at N. Philadelphia recorded a summer mean rainfall of 10 mm during 24 h and only one rainfall episode exceeding 40 mm during 24 h for the period 1970–1997. Also, the flash flooding episode of the 21st–22nd October 1994 in Attika has been attributed to extreme rainfall intensities reaching 42.7 mm during 24 h at N. Philadelphia rainfall monitoring station and having turnover time larger than 500 years (Mimikou and Koustoyannis 1995; Xanthopoulos et al. 1995).

Application of the Precis model at eastern Mediterranean predicted decrease in the mean monthly rainfall and in the number of days with rain for all months of the year at Athens during the period 2071–2100 (otroni et al. 2005). Also, results of the application of four regional climatic change models (HadCM2, Hadley Centre for Climate Prediction and Research; CCCma, Canadian Centre for Climate Modelling and Analysis; LMD, Laboratoire de Meteorologie, France;

and DKRZ, Deutsches Klimarechenzentrum, Germany) point to Athens as one of the localities in eastern Mediterranean experiencing the largest decrease in rainfall till 2050 (Feidas and Lalas 2000). Furthermore, there is expected higher frequency of droughts in localities that will exhibit the largest decrease in rainfall since the likelihood of drought days and their duration will increase; such indications are already apparent in Attika (Feidas and Lalas 2000). It seems, then, that Attika exhibits increased vulnerability to both droughts and floods due to climate change.

FLOODING MANAGEMENT

Under the name "Xenocrates," a general plan exists for the prevention, mitigation, and control of natural hazards including floods (Koutsoyiannis and Mimikou 1996). This plan constitutes a guide for collaboration and coordination of the authorities and public services organizations, which are involved in the control of emergency situations. The plan includes measures in both the central and local level. The authorities of the central level, i.e., ministries, are responsible for the general planning and coordination. The authorities of the local level (Prefectures, Municipalities, Communities) are responsible for the actions taken for prevention, e.g., inspection and maintenance of flood protection works, control (readiness both in equipment and personnel), and restitution (assessment, reporting, compensation) of damages.

Pre-emergency measures are focused mainly on the construction of flood protection works, such as levies and channel modifications, rather than forecasting and decision making. Attention has been given to flood forecasting for the protection of dams during construction and for the avoidance of dam failure, however (Koutsoyiannis and Mimikou1996). Yet, flood protection works in Attika have been repeatedly characterized as obsolete based on archaic data; inadequate to keep up with urbanization levels; faulty (positioned upstream instead of downstream), relocating flooding vulnerability to other localities within Attika; lacking proper operation and maintenance due to administrative confusion (Ministry of Planning, Public Works and Environment, Athens Water Supply and Sewerage Company (EYDAP SA), regional and local government bodies); fragmented aiming to cover urgent needs or to protect new road works; and, in some cases, incomplete (imikou 2005; Xanthopoulos et al. 1995; Greek GeoTechnical Champer 2004; Kitsos 2004). Experts characterize the storm drainage network as still primitive and call for a master plan pertaining to the flooding protection of the Attika basin, that would also integrate the climatic change challenge, and the establishment of permanent rainfall monitoring stations within the basin (imikou 2005; Xanthopoulos et al. 1995; Greek GeoTechnical Champer 2004).

Recurrent, contemporary flash floods at Attika are mainly due to unplanned, widespread, and rapid urbanization. The vulnerability of Attika to flash flooding is augmented because of an, as yet, spasmodic, un-coordinated, mainly reactive flood-prevention strategy, which does not take under consideration the anticipated increase in intensity and rapidity of rainfalls due to climate change at Attika basin. Management of flash floods in Attika Prefecture is further complicated due to large forest fires, such as the catastrophic forest fires of summer 2007 (BBC News 2007a, b; Kitsantonis 2007; Athens News Agency2007), which, among others, exacerbate erosion and surface runoff.

REFERENCES

1.   Athens News Agency. 2007. Nightmare of fire at Attika with fires at Kalivia and Imittos. 26/08/2007. http://www.in.gr/news/article. asp?lngEntityID=826554&lngDtrID=244 (in Greek).

2.   BBC News. 2007a. Greek forest fire close to Athens. 29/06/2007. http://news.bbc.co.uk/2/hi/europe/6252676.stm.

3.   BBC News. 2007b. Forest fire grips Athens suburbs. 17/08/2007. http://news.bbc.co.uk/2/hi/europe/6950662.stm.

4.   Bitsika, P. 2003. How concrete covered Attika. The lost streams and the floods that drown us. The Vima Newspaper 26/10/2003 (in Greek).

5.   Bitsika, P., and . Tratsa, 2002. Extreme phenomena knock our doors. The Vima Newspaper 24/11/2002 (in Greek).

6.   Dore MHI. Climate change and changes in global precipitation patterns: What do we know? Environmental International. 2005;31(8):1167–1181. doi: 10.1016/j.envint.2005.03.004.

7.   Feidas, H., and D. Lalas. 2000. *Climatic changes in the Mediterranean*. Technical report of the National Observatory

of Athens for the Ministry of PlanningPlanning, Works and Environment, Athens.

8.   Greek GeoTechnical Champer. 2004. Meeting for the flood prevention of Attica. Flood-prevention. Department of Professional Development, Office of Works Production, Athens, 2nd vember 2004 (in Greek). (http://portal.tee.gr/portal/page/portal/SCIENTIFIC_WORK/EKDILOSEIS_P/EPISTHMONIKES_EVENTS/ANTIPLHMMYRIKH%20PROSTASIA%20ATTIKHS/142).

9.   Greek National Statistics Service. 2001. Population census 2001. http://www.technowatch.aueb.gr/news/may99/esye.htm.

10.   Kitsantonis, N. 2007. As forest fires burn, suffocated Athens is outraged. International Herald Tribune 16/07/2007.http://www.iht.com/articles/2007/07/16/news/greece.php.

11.   Kitsos, G. 2004. Implemented and under implementation flood-prevention works in Attica the last five-years and repercussions from the new transport and other works. Meeting for the flood prevention of Attica. Flood-prevention. Department of Professional Development, Office of Works Production, Athens, 2nd ovember 2004 (in Greek). http://library.tee.gr/digital/m2022/m2022_kitsos.pdf.

12.   otroni, V., K. Lagouvardo, S. Mirasgenti, E. Georgopoulou, and G. Sarafidi. 2005. Localised climatic predictions for eastern Mediterranean: Results for the period 2071–2100. Climatic changes observatory, National Observatory of Athens (in Greek). http://www.teamfortheworld.org/docs/reports/eaa1.pdf.

13.   Koutsoyiannis, D., and M. Mimikou 1996. Management and prevention of crisis situations: floods, droughts and institutional aspects. Country Paper for Greece. 3rd EURAQUA Technical Review, Rome.

14.   Lasda, O. 2005. *Flooding in Attika: extreme meteorological phenomenon or result of urbanization?* BSc Thesis, Dept. of Environment, University of the Aegean, Greece (in Greek).

15.   imikou . Flooding status in Greece. Greece: Department of Water Resources and Environmental Engineering, School of Civil Engineering, National Technical University of Athens; 2005.

16.   Mimikou, M., and D. Koustoyannis. 1995. 1994 Extreme floods in Greece. Research Workshop on the Hydrometeorology, Impacts,

and Management of Extreme Floods, Perugia (Italy), November 1995.

17. Ministry of Planning, Public Works and Environment. 2005. Planning of the Metropolitan Regions for Sustainable Development (in Greek).http://www.minenv.gr/index.html.

18. Nicolaidou, M. and E. Hadjichristou. 1995. *Recording and assessment of flood damages in Greece and Cyprus*. BSc Thesis, School of Civil Engineering, National Technical University of Athens, Greece (in Greek).

19. Sioras P. Time trends of streams and basic parameters of flow at Attika basin.Greece: School of Rural and Surveying Engineering, National Technical University of Athens; 2003.

20. sakiris, G. 2004. The "flooding" of the Olympics construction works. The Apogevmatini Newspaper 31/10/2004 (in Greek).

21. Tsiourtis, N.X. 2002. *Greece-water resources planning and climate change adaptation*. Water, wetlands, and climate change building linkages for their integrated management. Mediterranean Regional Roundtable Athens, Greece, December 10–11, 2002. http://hydrogis.geology.upatras.gr/HYD/WATER_RESOURCES_GREECE.PDF.

22. World Health Organisation. 2005. EM-DAT: The OFDA/CRED International Disaster Database. Université Catholique de Louvain-Brussels-Belgium.http://www.em-dat.net/.

23. Xanthopoulos, Th., D. Christoulas, M. Mimikou, D. Koutsoyiannis, and M. Aftias. 1995. A strategy for the problem of floods in Athens. In: *Proceedings of the workshop for the flood protection of Athens*, Technical Chamber of Greece (in Greek).

The Traps Behind the Failure of Malpasset Arch Dam, France, in 1959

Pierre Duffaut

Expert in Geological Engineering, France

ABSTRACT

The case of the Malpasset arch dam failure in 1959 has been widely exposed in scientific and technical forums and papers. The focus here is on the many traps which have confused the whole chain of bodies and persons involved, owner, designer, geologist, contractor, up to the state management officers. When the first traps were hidden inside geology, many more appeared, as well geotechnical, technical, fortuitous, and administrative. In addition to such factual factors, human and organizational factors may be today easily identified, when none of them was yet suspected. Both dam safety and rock mechanics benefited from the studies done since the Malpasset case, most of them within one decade.

INTRODUCTION

On 2 December 1959, the failure of Malpasset dam (Fig. 1) was a prominent industrial catastrophe in France within the 20th century, only second by number of victims to a coal dust explosion in Courrières mine 53 years before. It was also a clap of thunder in the world dam community as never before any arch dam had failed, as André Coyne had pointed when opening a symposium on arch dams in 1957 as president of International Commission on Large Dams (ICOLD). It is well known that many more have been built since worldwide, and far higher, without any failure either. Many papers have described the Malpasset case, from the early studies to construction, operation and failure, the expert reports, the trial minutes and many lab and site investigations launched in order to understand what went wrong, ending with five papers published in 2010 in *Revue Française de Géotechnique* (Carrère, 2010, Duffaut, 2010, Goguel, 2010a, Goguel, 2010b and Habib, 2010).

Figure 1: Malpasset dam, left, at end of construction, summer 1954 (photo COB); right, soon after failure, end 1959 (photo Mary).

No surprise the first traps have been geological ones: for long, igneous and metamorphic rocks had been experienced as impervious enough for reservoirs and strong enough for dam foundations. Here they proved impervious, but failed as foundation. As the author began working in 1948 with EDF, the French authority for Electricity, in the Geology Department, he took part in the studies of many dam sites in France. He is now one of very few living geological engineers (if even anyone worldwide) to have worked in the field of dams at this time, when a great number of sites were investigated, in France and abroad. Neither EDF nor the author had been involved in Malpasset dam before the failure, while they immediately manifested their highest interest in the case. EDF was yet operating many arch dams, and many more were at construction or design stage. To the author, the case may look as a family affair, as his father Joseph Duffaut, had spent his whole career in dams. When Malpasset failed, the author was resident engineer on an arch dam construction site, just completed; his father was the first civil servant sent to the site by the government, the day after, as head of the Dam and Electricity department in the Ministry of Public Works; he then followed all the studies and trial sessions and the author could benefit his early pictures on site as well as his philosophy, "from father to son". After that his career was turned from Geological Engineering into Rock Mechanics.

One knows that most rare accidents derive from many wrong events together instead of only one; many more traps were to be soon discovered in addition to geological ones: geotechnical tests on site and in lab showed unsuspected and very poor properties; technical rules about uplift were not applied to thin dams! Two fortuitous events at the same time confused the local authorities, a worksite downstream and a flash flood. Last, and the more, no independent state control had ever been done on this public project, neither before nor during construction and operation. Since the mid-20th century, partly under pressure from the most hazardous industries, oil, aerospace, and nuclear activities, non-technical factors of accidents safety have been studied more and more and many scientists pointed that complexity is a hazard in itself: they showed how the weight of human and organizational factors could be heavier than factual and technical ones. The Malpasset case can bring them one more example.

The purpose of the paper is to explain how people in charge have been abused by so many traps, while what had been done there up to completion of the dam was correct within the practices of 1950s. So the responsibility of the catastrophe must be shared by many bodies and persons, the last one being a prefect, the local representative of the government, who did not know that a dangerous structure inside his territory was not managed by a competent enough staff. Many dams worldwide have been deeply modified along the years and a few have been put out of service when their responsible manager happened to discover they did not behave as safely as expected. So the memory must be saved of André Coyne and his dam engineering Bureau, (appearing below under acronym COB, Coyne and Bellier), always active today as Tractebel Engineering France, a member of GDF-SUEZ.

WHAT HAPPENED

About 15 km from Fréjus, an old Roman city on the Côte d'Azur, along Mediterranean Sea, at a place called Malpasset (a bad pass for passing people), a dam had been designed and built in 1950s to provide irrigation and drink water from Reyran, a very small river. Prof. G. Corroy of Marseilles University delivered the geological report, André Coyne and his Bureau COB designed the arch dam (Fig. 2 and Fig. 3) and supervised the whole construction works, made by contractor Ballot. None of them kept any mission from the owner after completion, in spite of the dam having never been formally checked; along the first years its filling up was prevented by lack of expropriation of a fluorite mine upstream of the dam, and later this reservoir was left unused as the water distribution network had not been completed. A geodesy company made four yearly measurements on about 30 targets (Fig. 4), but nobody interpreted the results of the year 1959, which had been transferred lately to the owner.

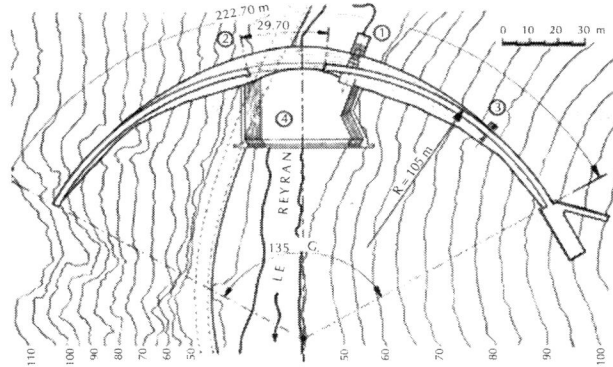

Figure 2: Site map, contours in meters over sea level (1: bottom gate; 2: surface weir; 3: water intake; 4: stilling basin); at right a gravity thrust block protected from water thrust by a wing wall counteracts the crest arch thrust.

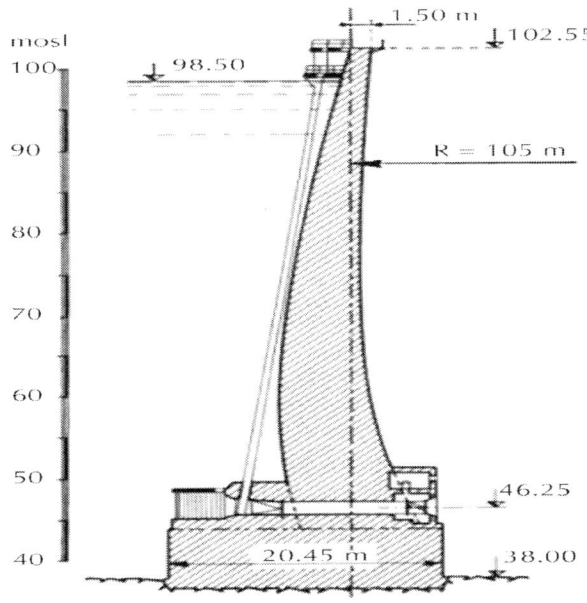

Figure 3: Highest dam cross section (mosl means meter over sea level). The support of the hollow valve and its control gate explains the widened foundation at this place only. Crest elevation: 102.55 m over sea level; spillway elevation: 100.4 m; normal operation level: 98.5 m; bedrock level: 38 m.

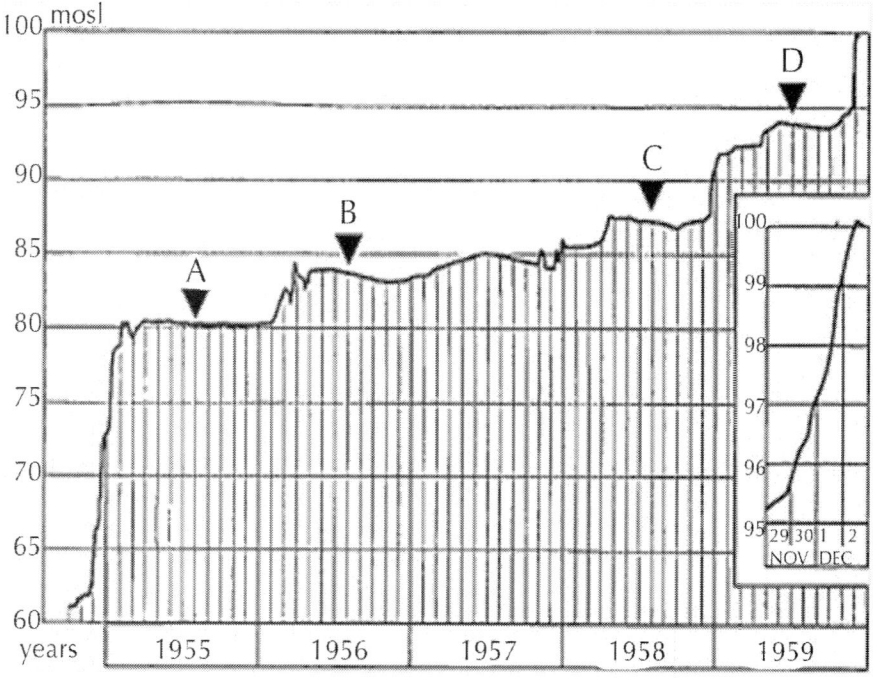

Figure 4: Graph of the reservoir level along years 1954–1959. The box magnifies the four latest days, and triangles mark the dates of geodesy measurements A–D (after Mary, 1968).

This year it rained a lot in autumn, resulting in the reservoir level increasing over any levels attained before. Contrary to previous years, the bottom gate was not opened to control the level because a motorway bridge was under construction 1 km downstream of the dam and nobody took account of any flow in the river. From November 30 to December 2, the rain intensity was such that the level rose 4.5 m in spite of the reservoir area increasing for each meter more (box in Fig. 4); the "normal" autumn rain had peaked as a sudden flash flood which was close to overflow the weir. In spite of a late opening of the bottom gate, the dam gave off at 23:11 and a huge wave wiped all structures along the valley, up to a small military airport on the seashore, making more than 400 casualties and a lot of destruction including all rail and roadways across the valley.

THE TRAPS

Geological Traps

The gross site of the reservoir was a narrow section through a small gneiss horst across a wide valley carved in coal measures, a rocky tract for easily damming a big reservoir. At first sight, the rock mass of this old metamorphic horst did not appear different from so many dam sites in other parts of France. After the failure, a flow of 50 million cubic meters of water have cleaned the slopes perfectly from any loose or even weathered material, the rock mass structure appeared very heterogeneous and crisscrossed by joints at any scale and in any direction as noticed by prominent geologist Jean Goguel (Goguel, 2010b) who surveyed the whole site soon after the failure. The geological history of the Estérel massif, now better known, may explain this peculiar structure which nobody had expected.

Only the failure daylighted two features of the rock mass which proved instrumental. A huge block of foundation rock was missing where was the left half of the dam arch, leaving an excavation in a dihedral form limited by two sub plane faces, always visible now (top of Fig. 5). Its downstream face is a true fault plane with a thin cover of crushed rock (fresh scratches on the surface proved the whole block had moved upwards). Its upstream face looks as a set of tears along two or more foliation surfaces, without any crushed rock. Neither the fault nor the foliation had been recognized before; the contours on Fig. 2 could have helped to infer the position of the fault, but its strike perpendicular to the valley axis and its dip about 45° upstream would have considered it as perfectly neutral with respect to the thrusts received from the dam; the continuity of rock foliation had not appeared either, within the so heterogeneous structure of the gneiss. So a geometrical trap was in place, waiting for a force susceptible to move the block, which will appear later.

Figure 5: The most conspicuous features of the site exposed after failure. Top, the "dihedral" excavation with half of the thrust block fallen after the flow (the exploratory adit at right was bored early after the failure to make jack tests in situ); below, a wide crevice is open just upstream of the concrete arch, wider at the base and closing more higher (photos Duffaut, 1960).

One may question why boreholes had not previously located the fault. First, most boreholes investigated the depth to the sound rock, under alluvium in the river bed and under any loose grounds along the valley slopes. Second, it is today difficult to recall that the technology of core recovery was not able to investigate such features. On many dam sites where contour lines accidents suggested a weak zone, their investigation used trenches instead of boreholes. The cleaning action of the flow made the fault path visible on the right bank and daylighted its cross section at either bank toe (Fig. 6).

Figure 6: Close view of a cross section of the main fault on right bank. The finely crushed borders of the fault zone are well visible, thickness about metric (photo Duffaut, 1960).

At rock matrix scale, some samples (Goguel, 2010b) revealed the rock close to the dihedral contained more *sericite* than elsewhere, a mica like mineral susceptible to increase the deformability and decrease the strength. He wrote: "I think … the failure is due to the poor mechanical strength of a gneiss which happens to contain dispersed sericite" (this observation was not followed with any tests of strength and deformability, and no rock block from the dihedral has been sampled to check this influence). Weathering of the gneiss has been pointed by some experts, but instead of any observation confirming its influence; the absence of any slope slide upstream of the dam during the fast drawdown and the perfect state of the roman aqueduct at mid-height proved the global strength of the slopes.

Geotechnical Traps

There had been no geotechnical investigations before the construction of the dam. Immediately after the failure, several studies were launched, at first on site, by seismic methods and jack tests, and in laboratories on samples taken from the site.

Refraction traverses showed a compact rock (velocity over 4000 m/s) below a shallow zone with velocity closer to 2500 m/s.

Petite Sismique (**Schneider, 1967):** A new short range seismic technology, easy to use on site, confirmed the high deformability of the rock in situ, together with the high compactness of the rock mass. The dynamic modulus derived from both deep refraction traverses and shallow *Petite sismique* was around 1500 MPa, a rather low figure for a dam foundation.

Jack tests (**Talobre, 1957**): EDF sent immediately a team to perform jack tests in a few small pits and galleries purposely bored. As very few such tests on dam sites were available for comparison, EDF ordered same tests be made on seven sites the same year and Malpasset provided by far the weaker results.

Later the practice became usual on most new sites, as shown in Fig. 7 which gathers the results for 17 sites and confirms the very high deformability of the rock mass, ten times less than that on most sites, hundred times less than that on best sites.

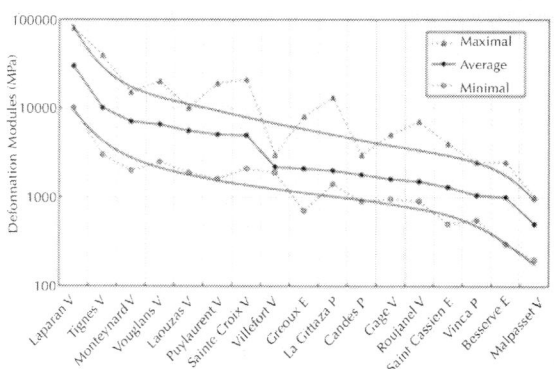

Figure 7: Deformation modulus measured on 17 dam sites (V: arch; P: concrete gravity; E: rockfill). Modulus scale in MPa, logarithmic; sites classified by decreasing modulus along three curves, maximal, average and minimal values (graph designed by B. Goguel, from EDF data).

Such low deformability was unsuspected and even the Saint Cassien dam site, close to Malpasset on the same rock type had provided results two times more.

Lab tests: Rock samples were sent to various labs, mainly École Polytechnique, Palaiseau (LMS, Laboratoire de Mécanique des solides), École des Mines, Paris-Fontainebleau, and École de Géologie, Nancy. The main set of tests results was described and discussed in a thesis work at LMS under supervision of Pierre Habib (Bernaix, 1967).

   Standard uniaxial compressive and tensile tests were performed on cylinders with diameters 10–60 mm and the same height to diameter ratio, 2.0. Strength values did not appear too low in average (58 MPa at dry state, and 42.5 MPa at saturated), but their *scatter* appeared by far wider than usual (coefficient of variation about 0.36), so providing many spots with high deformability. Taking various dimensions, a high scale effect was detected, as seen in Fig. 8. Systematic studies on various rock types showed that scatter and scale effect went together and provided a reliable fracturation criteria.

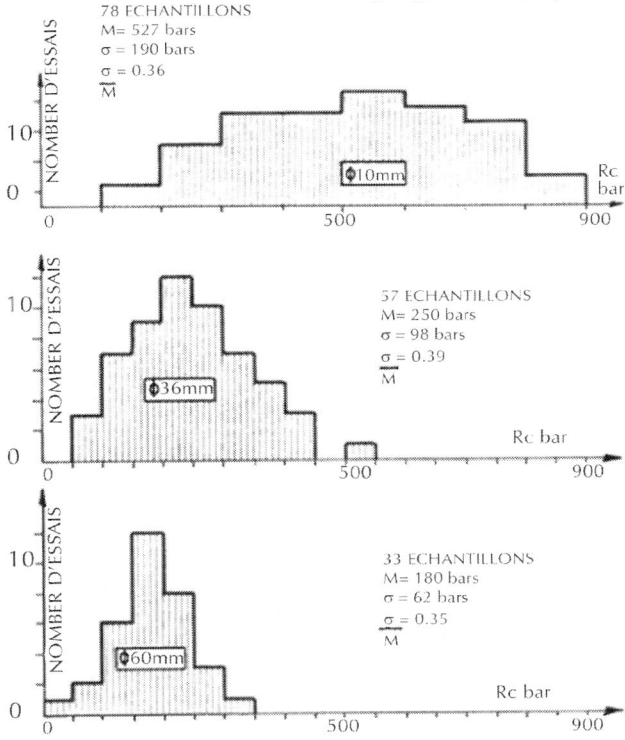

Figure 8: Distribution of unconfined compressive strength Rc in bar (=0.1 MPa) on dry samples (échantillons) collected on the left bank; three sets of cylinders, diameters 10, 36, and 60 mm (Bernaix, 1967). Comparison between three cases shows the scale effect: maximum strength occurs on smaller cylinders, where scatter also is the highest (vertical axis: number of samples; M: average value; : standard deviation).

Actually, all types of classical lab tests were performed, and one more: the permeability being very low, Habib (2010) proposed a new type of test to make measurements without any risk of error due to leaks along the envelope of the sample; thanks to a coaxial hole in the rock cylinder, a radial flow is generated from or to the hole depending if the fluid pressure is applied around or inside. When the flow is centrifuge, the rock is set in a tensile state which makes the permeability increase with the pressure, whatever the rock type; conversely a flow toward the hole creates a compressive state which does not alter the permeability of most rock types; but all samples from Malpasset showed a high sensitivity of the permeability, the more on the left bank: though rather low, the permeability decreased a lot under compression, due to closure of minute cracks. Not any other rock displayed such a behavior.

This unsuspected property, high sensitivity to stress, was then supposed to be the main cause of failure, as the load applied by the dam on the foundation rock induced a deep "underground dam" against which the uplift pressure could build more and more. Here was the force to move the dihedral. Actually, the high deformability was susceptible to play the same role: when the dam moves downstream under the water thrust, the rock upstream does not follow and a crack opens between the concrete and the rock as seen in Fig. 5 (below). The more deformable the rock mass is, the wider the crack opens, the deeper it extends, so increasing the height of the dam with a hydrostatic thrust increasing as the square of the height. Whichever mechanism prevails, the force on the dihedral is the same.

The anisotropy of the rock mass may provide another more mechanism: Maury (1970) investigated the stress bulb under a foundation which becomes thinner and deeper when the thrust is perpendicular to the stratification or schistosity; this influence was not well known at the time.

Technical Traps

After the failure of Bouzey small gravity dam (eastern France), Lévy (1895) showed that pressure from water seeping below and inside a structure plays as Archimede's thrust on buoying vessels and named it *sous-pression* (uplift). Most gravity dams were since preserved from it by relief holes; a so-called "drainage curtain" became a corollary of classical tightness curtains, but thin dams were thought immune thanks to the smaller area of their base. Malpasset opened the eyes of dam designers on uplift acting not only below and inside the structure, but also inside the rock mass downstream.

It became clear that any dam is a gravity dam, contrary to the usual classification of dams (Duffaut, 1992) provided enough ground mass be included in the gross resistant weight against the water thrust.

Incidental Traps

- **Money Inflation:** In 1950s, the money in France (and in many more Europe states) was inflating at a high rate. The sum allowed for the project was fixed, the owner was pressed to see the job finished in order to avoid the cost rise and he did not follow a recommendation of the geologist to make some more investigations. This monetary trap has been focused by Jean Goguel at the trial (Goguel, 2010a).

- **Bridge Worksite:** In 1959, a motorway was to be built from Aix-en-Provence to Nice which had to cross the Reyran river about 1 km downstream of the dam: earthworks had begun during summer months and the bridge worksite was glad to benefit a zero discharge in the river thanks to the dam; so the gate was kept close and the level rose higher than ever before, without any extra test or survey.

- **Flash Flood:** At the end of November, a flash flood occurred and the level rose dramatically (box in Fig. 4). One may notice such floods had been neglected from the design stage.

- **Geodesy Contract:** One could add the late delivery of the August 1959 measurements: the director of the company in charge of them was called to the trial for having delayed their delivery: he argued the length of calculations plus the summer vacations of his staff. Formally his contract did not ask for any interpretation or comparison with previous results. When the owner received the results (Fig. 9), nobody had paid attention and they were forwarded to the prefect to be included in the dam files.

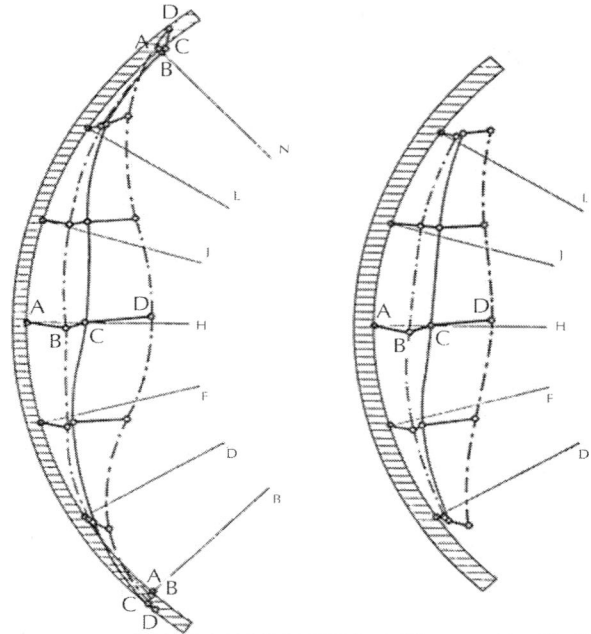

Figure 9: Displacements of the targets along arches at elevation 78 and 90 m (after Mary, 1968); bold letters refer to measurement dates (Fig. 4); other letters name the construction joints between monoliths; the scale applies to target displacements: segments CD, between measurements 1958–1959, show a general move toward the left bank, a fact any professional could have noticed as a change of behavior, long before the latest 5 m rise.

- **Cracks in the Stilling Basin:** No report was produced which confirmed the date, the location and the importance of appeared cracks a few days before the catastrophe in the reinforced concrete of the stilling basin without any schemes and photographs. One wonders that the guard was not specially auditioned on a material element susceptible to alert the staff in charge, at any level, on a disorder susceptible to be a sign of imminent danger. This element reflects the importance that each member of an organization is aware of its role in the safety of operation. If it is not an unfounded rumor, it is a grave failure of the organization, maybe a grave fault of the guard. At least, the concern of the staff in charge had led them to summon Coyne (and Ballot) at a close date, which has proved too late.

State Management Traps

The Var *département*, an administrative level of the State, was the owner, deprived of any dam specialist. It relied on the Engineer for design and for construction supervision but did not asked for any more mission after the end of construction; the formal *réception* (commissioning) did not wait for any test at full level, as was usual but not mandatory, by lack of water and because the property rights of the whole reservoir area had not been bought; the prefect signed it in order to permit full payment of contractors without asking the owner to perform any more surveys.

HUMAN AND ORGANIZATIONAL FACTORS

Lack of dialog with more than two persons (the Geologist sent reports but did not meet the Engineer on site). In France, Coyne used to work with EDF, a national company with many experimented engineers, and under permanent supervision of a corps of well-trained state engineers. Neither EDF nor this corps were involved in Malpasset.

One may wonder administrative borders be tight against technical information exchange: along the same years a bigger arch dam, Bimont, was being built within 100 km of Malpasset for drink water of Marseilles under the next *département* Bouches-du-Rhône as owner, but not any link appeared between the teams in charge of those dams in spite of both the geologist and the engineer being the same.

And the more one must notice the lack of any control by a third party (which was soon created and made mandatory for all dams over 15 m height).

OTHER CATASTROPHES

Many comparisons may be made with a series of catastrophes within the 20th century and up to now:

- coal-dust explosion in the Courrières mine in France in 1906, when nobody could think a coal-dust explosion could reach so far;

- drowning of the British liner Titanic in 1912 after collision with an iceberg: from the officers and all passengers on board, to the general public worldwide, the liner was thought of as non-submersible and the staff denied the cables received about the drift of icebergs;

- explosion of the English dirigible R101 (1930, near Beauvais, France) during its inaugural flight, operated before completion of the convincing steps of the experimental stage, to fulfill political ambitions: a fault shared with Malpasset and Vajont (see below);

- deadly Aberfan debris flow in South Wales in 1966: in spite of many small shallow slides on sterile coal mine heaps, the height has been increased higher than experienced before (Duffaut, 1982);

- explosion of the American Space Shuttle Challenger (1986) after many warnings about failed joints in the solid rocket booster and the decision to maintain the launch to meet political requirements of NASA (aimed to obtaining next budgets);

- crash of flight AF447 (offshore Brasil, 2009), the staff of which was caught in a tropical tempest and deprived from data from frozen Pitot tubes;

- Fukushima tsunami induced nuclear accident (2011, Japan), where the local staff was left without any means of action when both water cooling of the reactors and electric power were put out of service by a wave higher than supposed.

Soon after Malpasset, another dam catastrophe occurred in the Dolomites, Northeastern Italy: the fall of Monte Toc slope into the Vajont reservoir on 9 October 1963. It displaced the water from the reservoir, which swept 2000 people in the valley downstream: as the slope was moving slowly along two years, the engineers thank they could control the slide through management of the reservoir level, ignoring the lesson of the celebrated Goldau slide, Switzerland, 200 years before, which had suddenly accelerated and destructed the city (Heim, 1932 and Erismann and Ebele, 2001). Leopold Müller who was in charge of the geotechnical studies focused on the reinforcement of the rock mass through many rock anchors on both banks (which actually proved efficient: Leonards (1987) stated that the Vajont dam withstood a load eight times greater than it was designed to bear). The actual trap was the confidence in the management of the slide in front of a transfer of property from a local company to a State one.

HUMAN AND ORGANIZATIONAL FACTORS OF CATASTROPHES: "NORMAL ACCIDENT" THEORIES

In addition to the lack of civil Rock Mechanics, which was to be derived from the studies following the catastrophe, one must notice that social research on major accidents was also in infancy in the fifties: after works on accidents by Patrick Lagadec in France, Charles Perrow in the US, James Reason in UK and many other since, under pressure from high hazardous activities, it is now well understood that nature and industry build together so complex systems than nobody can any longer master all hazardous interactions inside them.

In France, Lagadec (1979) introduced the concept of Major Technological Hazard and performed reviews of many major accidents. In his famous book, "Normal accidents", Perrow (1984) wrote after the Three Mile Island nuclear accident: "we might stop blaming the wrong people and the wrong factors"; he stated that in complex systems, "multiple and unexpected interactions of failures are inevitable": the accident becomes "normal"! The aerospace industry, NASA at the first place, ordered many studies which benefited to all most hazardous industries, nuclear energy to begin with. These studies have quickly highlighted the importance not only of human functioning but also the influence of organizations as outlined by James Reason "we cannot change the human conditions, but we can change the conditions under which people work" (Reason, 1990).

CONCLUSIONS

The geology set the first traps; the mechanical behavior of the rock aggravated the dangerous forces; the practice of drainage only was of rule under thick dams; two fortuitous circumstances, the construction of a bridge and a flash flood, conjugated; all of those traps were in a way preparatory causes. Money inflation, lack of any state control, blindness in front of alarms, and absence of any qualified staff completed the scenery. One should stress as well the technical isolation of André Coyne, instead of the high level of implication of engineers of both EDF and the state in hydro-dams inside France.

It was highly uneasy, either at the trial, some years after, or, some decades later at Purdue University (Leonards, 1987), and it is yet today uneasy too, half a century later, to discuss how engineers performed in the early fifties; it may look easy to charge them with outrageous transgressions of elementary rules of art, when no such rules did exist at the time: most of the rules enforced today have been derived from the results of the Malpasset case history; within a few years, many had yet become evident. Geological materials are opaque, we cannot see through, so it may be compared with a lock, the mechanism of which is purposely hidden behind a steel plate; in most geotechnical problems the thrust to turn the key is provided by groundwater. While geology may be investigated as much as needed, future events are not predictable, from rain and flood to earthquakes and tsunamis, to worksites, and even societal movements.

For sure the catastrophe has brought many useful teachings: the way dam sites were investigated before construction and the way dams were managed during operation has been since deeply changed worldwide, but my purpose here is to recall what has been done before construction was "normal" at the time, while what has been done after was not.

ACKNOWLEDGEMENTS

Colleagues of both French committees for Dams and Reservoirs and for Rock Mechanics, Jean-Louis Bordes, Bernard Goguel, Pierre Habib, and younger doctorant Justin Larouzée. Some more valuable information may be found in Bellier (1967), Londe (1987), and Duffaut (2011).

REFERENCES

1.   Bellier, 1967 J. Bellier Le barrage de Malpasset Travaux, Paris (1967), pp. 3–23

2.   Bernaix, 1967, J. Bernaix, **Étude** géotechnique de la roche de Malpasset Dunod, Paris (1967)

3.   Carrère, 2010, A. Carrère Les leçons de Malpasset, leur application aux projets de barrages d'aujourd'hui Revue française de Géotechnique, 131/132 (2010), pp. 37–51

4.   Duffaut, 1982, P. Duffaut, La rupture du terril d'Aberfan (Pays de Galles), 21 Octobe1966, Industrie Minérale (1982), pp. 413–418

5.   Duffaut, 1992, P. Duffaut Any geotechnical engineer has to face rockmass water, Coll. Ascona EPF Lausanne (1992)

6.   Duffaut, 2010, P. Duffaut, Malpasset, la seule rupture totale d'un barrage-voûte Revue française de Géotechnique (131/132) (2010), pp. 5–18

7.   Duffaut, 2011, P. Duffaut, What modern rock mechanics owe to the Malpasset Arcdam failure Proceedings of International ISRM Congress, CRC Press, Leiden (2011), p. 701

8.   Erismann and Ebele, 2001, T. Erismann, G. Ebele, Dynamics of rockslides and rockfalls Springer, Berlin (2001)

9.   Goguel, 2010a, B. Goguel Avant-propos au rapport géologique Malpasset de Jean Goguel Revue française de Géotechnique, 131/132 (2010), pp. 23–24

10. Goguel, 2010b, J. Goguel Rapport géologique Malpasset Revue française de Géotechnique, 131/132 (2010), pp. 25–36

11. Habib, 2010, P. Habib La fissuration des gneiss de MalpassetRevue française de Géotechnique, 131/132 (2010), pp. 19–22

12.   Heim, 1932, A. Heim, Bergsturz und Menschenleben Fretz and Wasmuth, Zurich (1932)

13. Lagadec, 1979, P. Lagadec Le défi du risque technologique majeur Futuribles, 28 (1979), pp. 11–34

14. Leonards, 1987, G.A. Leonards Proceedings of the International colloquium on Dam failures, Purdue University, Lafayette, Indiana (1987)

15. Lévy, 1895, M. Lévy Quelques considérations sur la construction des grands barrages Comptes rendus de l'Académie des Sciences, Paris, 121 (1895), pp. 288–300

16. Londe, 1987,P. Londe The Malpasset dam failure Engineering Geology, 24 (1987), pp. 295–329

17. Mary, 1968, M. Mary Barrages-voûtes, historique, accidents et incidents Dunod, Paris (1968)

18. Maury, 1970, V. Maury Mécanique des milieux stratifies Dunod, Paris (1970)

19. Perrow, 1984, C. Perrow Normal accidents: living with high risk technologies Princeton University Press, New York, NJ (1984)

20. Reason, 1990, J. Reason Human error Cambridge University Press, Cambridge (1990)

21. Schneider, 1967, B. Schneider Moyens nouveaux de reconnaissance des massifs rocheux Annales Institut technique du Bâtiment et des Travaux Publics, 235/236 (1967), pp. 1055–1093

22. Talobre, 1957, J. Talobre La Mécanique des roches appliquée aux Travaux publics Dunod, Paris (1957)

Treatment Design of Geological Defects in Dam Foundation of Jinping I Hydropower Station

Shengwu Song, Xuemin Feng, Hongling Rao, and Hanhuai Zheng

Chengdu Hydroelectric Investigation and Design Institute, China Hydropower Engineering Consulting Group Corporation, Chengdu 610072, China

ABSTRACT

Jinping I hydropower station is one of the most challenging projects in China due to its highest arch dam and complex geological conditions for construction. After geological investigation into the dam foundation, a few large-scale weak discontinuities are observed. The rock masses in the left dam foundation are intensively unloaded, approximately to the depth of 150–300 m. These serious geological defects lead to a geological asymmetry on the left and right banks, and thus some major difficulties of dam construction are encountered. In this paper, the influences of geological defects on the project are analyzed, followed by the concepts and methods of treatment design. Based on the analysis, the treatment methods of the weak rock masses and discontinuities are carefully determined, including the concrete cushion, concrete replacement grids, and consolidation grouting. They work together to enhance the strength and integrity of the dam foundation. Evaluations and calibrations through geo-mechanical model tests in combination with field monitoring results in early impoundment period show that the arch dam and its foundation are roughly stable, suggesting that the treatment designs are reasonable and effective. The proposed treatment methods and concepts in the context can be helpful for similar complex rock projects.

INTRODUCTION

Jinping I hydropower station is located in Liangshan Yi Autonomous Prefecture, Sichuan Province in Southwest China. It is built on the Yalong River as a controlling cascade hydropower station in the middle and lower reaches, with power generation as its major function. It has annual regulating capacity of water resources and provides remarkable compensation benefits for the lower cascade hydropower stations. The total reservoir capacity is 7.76×10^9 m^3 at a normal water level of 1880 m, and the installed capacity is 3600 MW. The height of concrete double-curvature arch dam is 305 m with the bottom foundation elevation of 1580 m and dam crest elevation of 1885 m, which is the highest under construction in the world.

The hydropower station is built in the transitional slope zone between Qinghai-Tibet Plateau and Sichuan Basin. In the project area, the river valley is sharply incised with high steep slopes on both banks. Large-scale tectonic faults with strikes of river direction are developed and rock masses in left bank are intensively unloaded with depth of 150–300 m (Liu et al., 2010 and Song et al., 2011), leading to a geological asymmetry on the left and right banks. For these geological defects existing, the rock mass quality in the studied area basically cannot meet the requirements of the dam foundation construction.

Consequently, the arch dam project is regarded as one of the most challenging projects with complex geological conditions and difficult foundation treatments (Liu et al., 2004 and Fan et al., 2012).

The design of Jinping I hydropower station was completed in 2003. After the excavation of dam foundation, the subsequent concrete placing of the dam started in 2009. At the end of 2012, the diversion tunnels were plugged and early impoundment of the reservoir began. At present, the concrete arch dam is constructed up to a height of 280 m, and most of the foundation treatments have been implemented. According to the schedule, the dam will be impounded and put into power generation in August 2013. In this paper, the analytical methods, design concepts, treatment measures and technical features of strengthening the geological defects in the dam foundation will be presented.

GEOLOGICAL SETTINGS OF DAM FOUNDATION

In the project region, the Yalong River flows along the direction of N25°E, with water level of 1635 m in dry season. The top elevation of the valley mountains reaches up to 3200–3600 m, and relative height difference of topography is 1500–2000 m. High and steep slopes stand in precipitous gorges and V-shaped valleys are sharply incised, with a declination of 40°–90°.

The outcropping strata in project region are group T_{2-3z}, the middle–upper group of Triassic Zagunao formation, which consists of the core parts of the dam foundation and southeast branch (normal branch) of the close inversed Santan Syncline. The rocks in the region can be classified into three groups according to their lithology. The first group is green schist (T_{2-3z}^{1}), which is buried in depth under the riverbed and does not outcrop on the ground. The second one is marble of green schist interbeds (T_{2-3z}^{2}). It consists of eight layers observed in both banks of the river with thickness of about 600 m. This rock group contributes to the host rocks of the dam foundation, and interlayer belts develop in the green schist rock masses. The third one is metasandstone and slate (T_{2-3z}^{2}) consisting of six layers, with overburden thickness of about

400 m. It is formed in the core strata of Santan Syncline and can be found in upper left abutment. Rock mass integrity of this group is not good as expected because of its lithology and formation. The strikes of those strata are almost consistent with river flowing direction, dipping toward the left bank with an inclination of about 40°. Therefore, a typical longitudinal valley is formed, with consequent slopes in right bank and inverse slopes in left bank (Fig. 1).

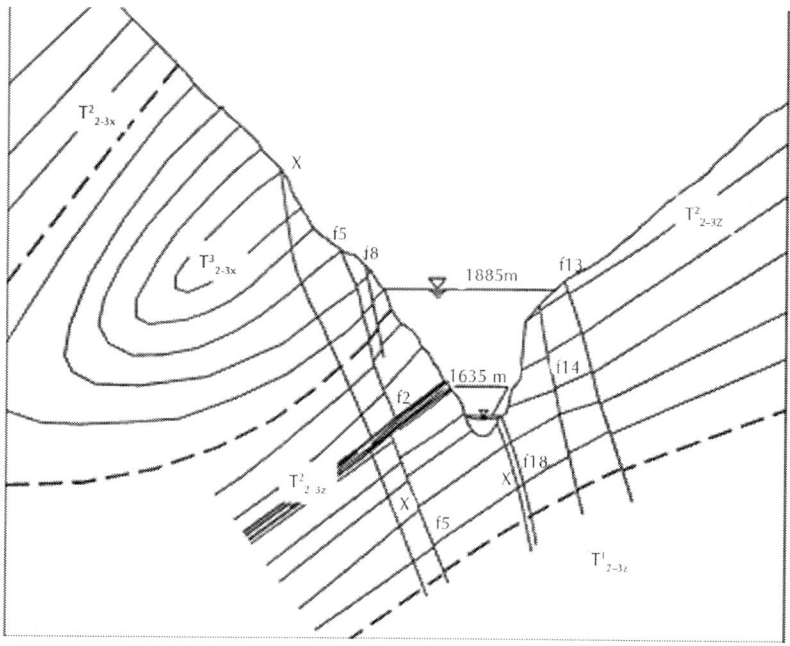

Figure 1: Sectional diagram of geological structures across the valley in the project area.

Weak discontinuities, mainly in the presence of faults, are developed in the dam site, with predominating strikes of NE–NNE and dip of SE. Some large-scale weak discontinuities are of faults f5, f8, f2 and lamprophyre dyke X in left bank, faults f13, f14 in right bank, and fault f18 lying obliquely in the riverbed, which definitely control the stability of the project (Fig. 2).

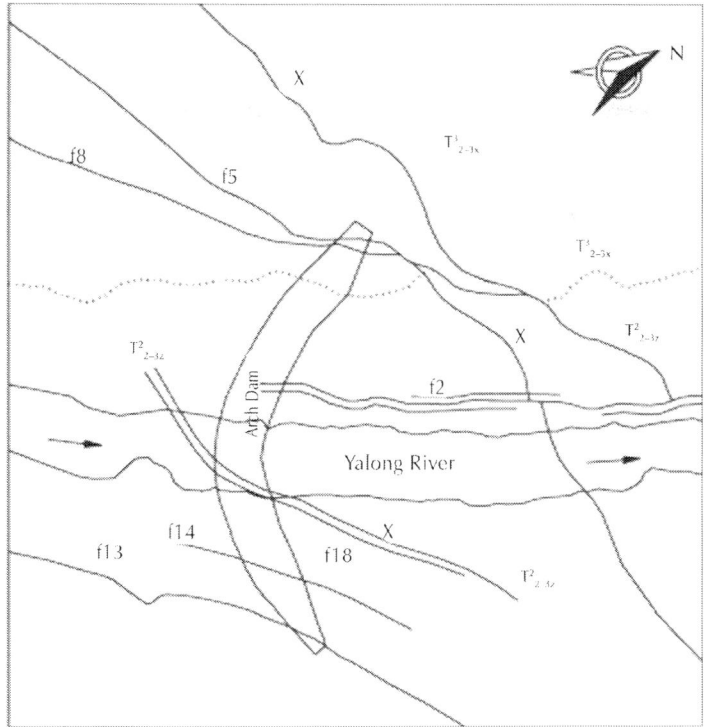

Figure 2: Planar diagram of geological structures in the project area.

Faults f5, f8 and lamprophyre dyke X are observed outcropping in the left bank, and the width of crushed zones in the faults is basically 2–8 m. These discontinuities are mainly composed of loose tectonic breccia, mylonite and strongly weathered rocks which are easy to dissolve in water. Owing to the unloading of valley slopes, the rock masses intersected by these weak discontinuities above the elevation of 1 680 m are intensively fractured and loosened. Fault f2 outcrops in the lower slope in the left bank, consisting of four or five displacing interlayer belts with broken green schist rocks. Faults f13 and f14 develop in the right bank, with breccia and mylonite as their main composition, and mud with a thickness of several centimeters can be found in the faults. Fault f18 lies in the riverbed oblique to the right bank, and it is composed of lamprophyre dyke X, strongly weathered and susceptible to water (Table 1).

Table 1: Features of main weak discontinuities (faults) in the project region

Faults	Attitude	Width of crushed belt (m)	Length of crushed belt (km)	Location
f5	130°∠70°	2.0–8.0	>1.8	Left bank slope
f8	125°∠75°	1.0–2.0	>1.4	
f2	290°∠40°	8.0–10.0	>1.0	
X	153°∠68°	2.0–3.0	>1.0	
f13	148°∠70°	1.0–2.0	>1.0	Right bank slope
f14	150°∠75°	0.5–1.0	0.5–0.7	
f18	165°∠75°	2.0–3.0	>1.0	Right riverbed

The rock masses are intensively unloaded in the slopes of the valley. In the right bank, the depth of unloading generally reaches 30–50 m, which is within a normal unloading scope. But in the left bank, the unloading depth of rock masses is considerably large due to the impacts of geological structures and rock lithology. In middle and lower slopes, this unloading depth of marble is 150–200 m, and that of sandy slate in middle and upper slopes can reach 200–300 m and even extend 500 m along the river side. Relaxed tensile fissures resultant from original joints can be frequently encountered in the unloaded rock masses, with apertures of 10–20 cm. This special geological phenomenon is rarely reported and thereafter is named as deep unloading (Qi et al., 2004 and Song et al., 2011). The unique unloading feature in this region is the primary factor leading to the geological asymmetry of both banks. According to geological investigation in the dam site, the rock quality is classified as listed in Table 2.

Table 2: Rock mass quality classification and physico-mechanical parameters in the dam site

Grade	Main geological feature	R_c (MPa)	V_p (m/s)	E_o (GPa)
II	Fresh intact marble	60–75	>5500	20–32
III_1	Fresh marble with green schist interbed	60–75	4500–5500	10–14
III_2	Slightly unloaded marble and sandy slate	40–75	3800–4800	3–10
IV	Strongly unloaded and deep unloaded marble and sandy slate	–	<3500	1–4
V	Broken belts of faults and crushed displacing interlayer belts	–	–	<1

Note: R_c is the uniaxial compressive strength, V_p is the acoustic P-wave velocity, and E_o is the deformation modulus. And the classification standard is roughly based on "Code for hydropower engineering geological investigation" (GB 50287-2006).

MAJOR GEOLOGICAL DEFECTS IN DAM FOUNDATION AND INFLUENCES

It is previously mentioned that the geological condition in the dam site is significantly asymmetric, relatively good in the right bank but unfavorable in the left bank. Thus a series of technical challenges for the dam foundation treatment are encountered. Through detailed analysis, the geological defects in dam foundation can be classified into two categories. One is the rock mass of grades III_2 and IV affected by unloading and ripping in dam foundation and resistance blocks of left bank, the other is large-scale discontinuities developed in the dam site (Liu et al., 2010).

According to the geological investigations and field excavations, rock masses of grades II and III_1 in this region account for 89.8% of rock masses of the right bank in total. In the left dam foundation, the rock mass quality of the upper part of slope is significantly different from that of lower part. Below the elevation of 1780 m, the foundation surface is dominated by rock masses of grades II, III_1, about 87.9%; while above that level, they are mainly of rock masses of grades IV–V, about 72.7%, and no rock masses of grade II can be found and that of grade III_1 only accounts for 15.1% of rock masses in total (Table 3 and Fig. 3).

Table 3: Rock mass quality assessment of different parts in the dam foundation surface.

Location	Percentage of outcropping area (%)				
	II	III_1	III_2	IV	V
Right bank	73.0	16.8	8.7	0.7	0.8
River bed	39.4	60.6	–	–	–
Left bank (above EL.1780 m)	–	15.1	23.2	17.2	55.5
Left bank (below EL.1780 m)	37.2	50.7	3.0	5.1	4.0

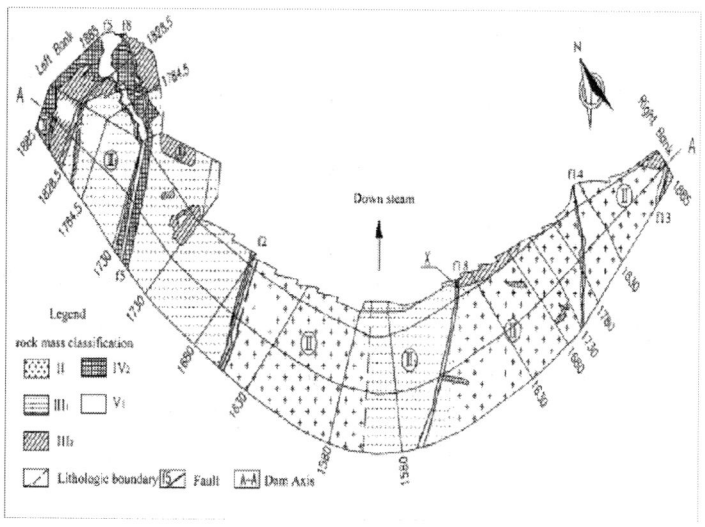

Figure 3: Distribution diagram of rock mass quality in dam foundation surface (unit: m).

It is noted that rock masses of grades III_2, IV and V cannot be utilized as the foundation host rocks directly due to their unfavorable effects on stability of the dam foundation. After effective engineering treatments, rock masses of grade III_2 can be considered as the foundation host rocks. However, special treatments are desired for the rock masses of grades IV and V in the dam foundation.

In the left valley slopes, the faults f5 and f8 featured with shallow overburden depth are observed, where fractured rock masses are widely distributed. The strongly unloaded rock masses outside the faults in combination of the faults exert adverse influence on the stability of the dam. Meanwhile, it is clear that the faults f5 and f8 attribute to the side boundary of potential sliding surfaces in the left abutment. Lamprophyre dyke X, buried in depth inside the slope, is strongly weathered, and its strike is almost perpendicular to the direction of the dam thrust, which also imposes unfavorable influence on the stability of the dam foundation. The lamprophyre dyke X similarly constitutes a potential boundary surface of certain sliding block in the left abutment (Song et al., 2010 and Wang et al., 2012).

In the vicinity of riverbed, the faults f2 and f18 lie in the lower part of the dam foundation (i.e. elevation 1600–1680 m). Moreover, they are located in the main bearing zones of the dam foundation, and thus have major effects on the stress distribution and seepage stability of the dam foundation. The fault f2 composes a potential bottom surface of certain sliding blocks in the left abutment, and therefore it is unfavorable for the anti-sliding stability of the dam.

In the right valley slopes, the faults f13 and f14 outcrop in the upper part of dam foundation. For this reason, ground water signs show the layer has good permeability. It is interesting to note that the faults are interconnected with the reservoir in the upstream, and in the downstream they pass through the underground powerhouse area. In this way, the faults impose adverse influences on the stress distribution, transferring of the arch thrust, seepage stability and surrounding rock mass stability of underground powerhouses in the right abutment. Similarly, the faults f13 and f14 also constitute the potential boundary surfaces of sliding blocks in the right abutment.

DESIGN OF TREATMENT FOR THE GEOLOGICAL DEFECTS

Concepts of Treatment Design

Based on the disclosed geological conditions in the project region, effective engineering measures should be adopted to strengthen these defects and to analyze the influence of geological defects on the project stability. In the view of technical feasibility, concepts of treatment design for the geological defects in different locations in the dam foundation should be carefully proposed.

1. The strikes of large-scale faults f5 and f8 are almost in parallel with the valley slope surface. The mechanical properties of those faults are very poor, thus excavation of the upper shallow parts of the faults and the slope rock masses outside them should be considered to improve the anti-sliding stability of the dam.

2. The deep lamprophyre dyke X and the lower part of fault f5 are composed of very weak rock masses, which have a fairly wide influential zones, both of them dip toward the river. Local excavation in grids can be adopted, and after excavation it can be backfilled by concrete or grouted directly, in order to improve the shear-resistance capability and the integrity of rock masses.

3. Fault f2 outcrops in the lower part of the dam foundation and it extrudes the displacing belts and fault f18 that lies in the right of the dam foundation. The shallow parts are considered to be excavated after slots cutting in the foundation surface, and then concrete backfill can be implemented together with flushing and grouting in certain deep areas, so that the bearing capacity and anti-seepage properties of rock masses in the vicinity of these faults can be improved.

4. Faults f13 and f14 are interconnected with the reservoir, and they have great impacts on stress distribution imposed on the dam foundation. Therefore, local places must be excavated and replaced with concrete. In addition, deep anti-seepage system and grouting are needed to reduce the negative influences of

these faults on the stability of the right abutment and underground powerhouses.

5. In the left dam foundation, the sandy slate above elevation 1800 m and the marble in the elevations of 1730–1800 m are quite fractured and the mechanical properties are considerably poor, which are directly associated with the faults f5, f8 and dyke X. Therefore, a massive concrete replacement cushion for those strata is needed; meanwhile, effective grouting will be implemented to improve the resistant capability of host rocks in this area.

6. Drainage structures are considered in the downstream of the dam foundation, where anti-seepage curtains and resistance blocks are adopted. \

Main Schemes of Treatment Design

Excavation Design of Abutments

The slopes in the right bank are mainly composed of marble that has relatively good mechanical property. The degrees of weathering and unloading depth of the marble formation are basically in a normal range, and the unloading depth is generally less than 50 m. The rocks in the depth below 50 m are almost fresh and classified into grade II or III_1, a good choice for host rocks of the dam foundation. In this way, the scope of slope excavation is limited within the boundaries of the weathered and unloading belts in the right abutment.

The slopes in the left bank are roughly composed of sandy slate and marble. The rocks above the elevation 1750 m are intensively unloaded with depth more than 200 m, most of which are grades III_2 and IV and cannot be directly utilized as dam foundation surface. Particularly, in the elevation 1750–1730 m, fault f5 is in the vicinity of the arch abutment. The rocks around it cannot be considered as the dam foundation either for too small thickness of the layer on one hand, and on the other hand, excess excavation of the abutment will lead to large deformation of the arch dam and heterogeneous distribution of stresses in the dam (Xiang and Rao, 2008 and Song et al., 2010). Therefore, the faults f5 and f8 and rocks around the faults (above elevation 1730 m) will be excavated in the left abutment.

Design Schemes of Concrete Cushion System for Weak Rocks in the Foundation Surface

The concrete cushion in the left abutment is constructed in the elevations of 1730–1885 m. The thickness of the concrete cushion at different elevations is required according to the arch dam shape and the excavation scope of the spandrel groove.

In order to design an optimized width of the concrete cushion foundation, sensitivity analyses by finite element method (FEM) at the representative elevations of 1750 m and 1830 m have been conducted. Six schemes with different foundation widths in the contact surface of the cushion and bedrock are selected. The ratio of the foundation width to the arch abutment thickness at the same elevation is 1.4, 1.8, 2.0, 2.2, 2.4 and 2.6, respectively. The displacements of the arch abutment in the direction of arch thrust are adopted as evaluation index. The results indicate that, the displacements of the arch abutment at the elevation 1750 m, with values from 31 mm to 34 mm, tend to be stable when the width ratio reaches 1.8. While at the elevation of 1830 m, the relative ratio is 2.2, with displacement of approximately 15 mm. Therefore, the values of 1.8 and 2.2 are selected as the optimized ratios in the design of the width of the cushion foundation at the two elevations, which can be also referred to as a design principle at other elevations.

For the concrete cushion design, four schemes at the elevation of 1830 m have been implemented by two-dimensional (2D) FEM. Of these schemes, the width of the cushion at the side of the arch abutment is set to be 1.0, 1.5 and 2.0 times the arch abutment thickness, respectively. And different cases of slope excavation outlines of the spandrel groove are considered. Results show that the displacements of the arch abutment in the direction of river flow and the stresses in the cushion differ slightly with the width ratios of 1.5 and 2.0. The values of displacements and stresses are 18–24 mm and 1.0–1.5 MPa, respectively. However, the displacement is up to 44 mm and the principal compressive stress is up to 3.2 MPa when the ratio is 1.0. In consideration of both safety and economy, the width of the concrete is designed at 1.5 times the arch abutment thickness. The final layout of the concrete cushion is shown in Fig. 4, with average width and average thickness of 61.2 m and 49.8 m, respectively. The total concrete volume is 560.2×10^3 m^3.

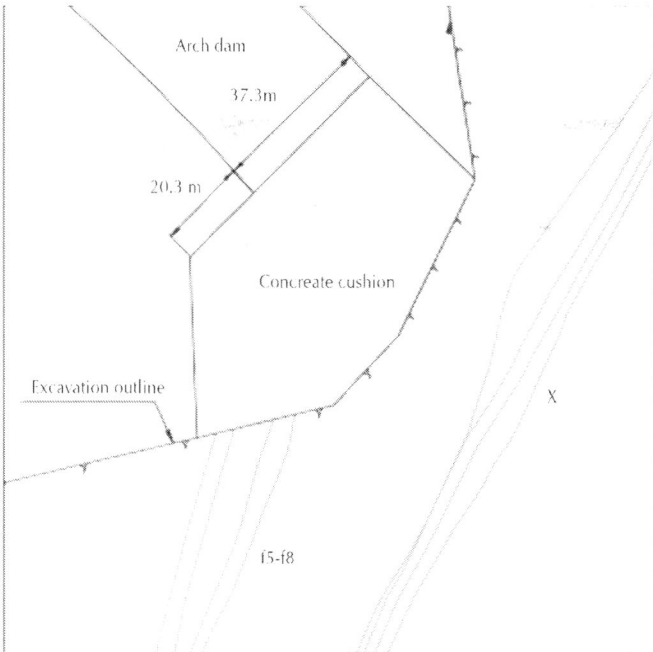

Figure 4: Layout and dimensions of the concrete cushion at the elevation 1830 m.

In the cushion foundation, the rocks on the surface are unloaded, relaxed and weak. In order to improve the force-transfer effectiveness of the arch dam and the stress distribution in the rocks, five concrete replacement galleries are designed in the cushion foundation: one at the elevation of 1829 m, two at 1785 m and two at 1730 m, with sectional dimensions of 9 m × 12 m (width × height). These force-transfer shear galleries pass through the lamprophyre dyke X are situated in the intact rock masses, which are helpful for providing additional anti-shear capability of lamprophyre dyke X.

Treatment Design of Concrete Replacement Grids for the Fault F5 and Lamprophyre Dyke X

In the left abutment, concrete replacement grids are adopted to treat the fault f5 and lamprophyre dyke X. Firstly, for fault f5 below the cushion foundation after excavation, two concrete replacement galleries at the elevations of 1730 m and 1670 m will be constructed along the strike of the fault. The height of the two concrete replacement galleries is 10 m, but the width varies according to the thickness of the crushed zones of the faults, basically 9 m in average. At the two elevations, four inclined concrete replacement shafts are employed along the fault plane, with a space of 30–35 m and a width of 15 m. The first shaft on the upstream also acts as an anti-seepage shaft of the dam curtains for fault f5, with a smaller width of 10 m (Fig. 5).

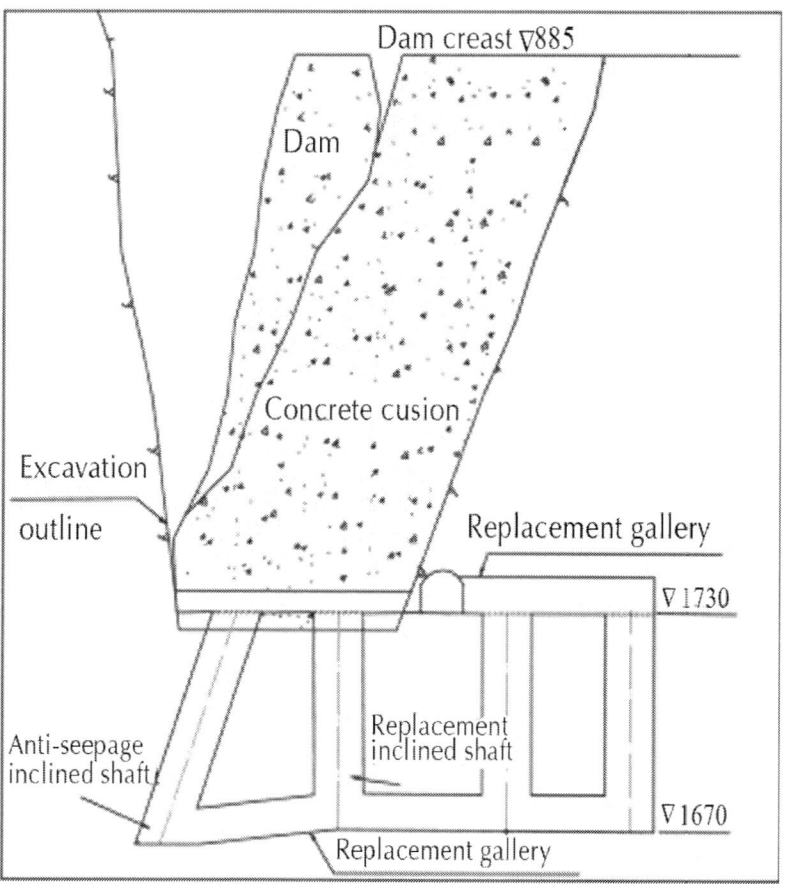

Figure 5: Schematic diagram of the concrete replacement galleries and shafts for fault f5 in the left abutment (unit: m).

Secondly, for the treatment of lamprophyre dyke X, three concrete replacement galleries at the elevations of 1829 m, 1785 m and 1730 m are considered with sectional dimensions of 9 m × 12 m (width × height), which are intersected with the above-mentioned force-transfer shear galleries. In the elevations of 1829–1785 m, an inclined replacement shaft is implemented, together with the anti-seepage curtains of the dam. In the elevations of 1785–1730 m, four inclined concrete replacement shafts will be constructed, with a space of 31 m and a width of 7 m (Fig. 6).

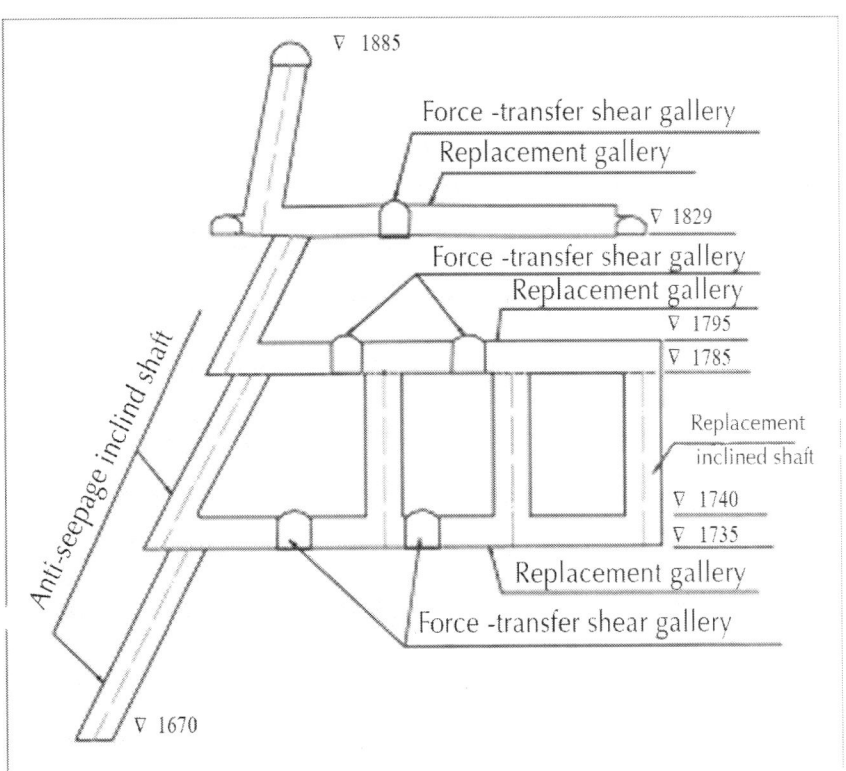

Figure 6: Schematic diagram of the concrete replacement galleries and shafts for lamprophyre dyke X in the left abutment (unit: m).

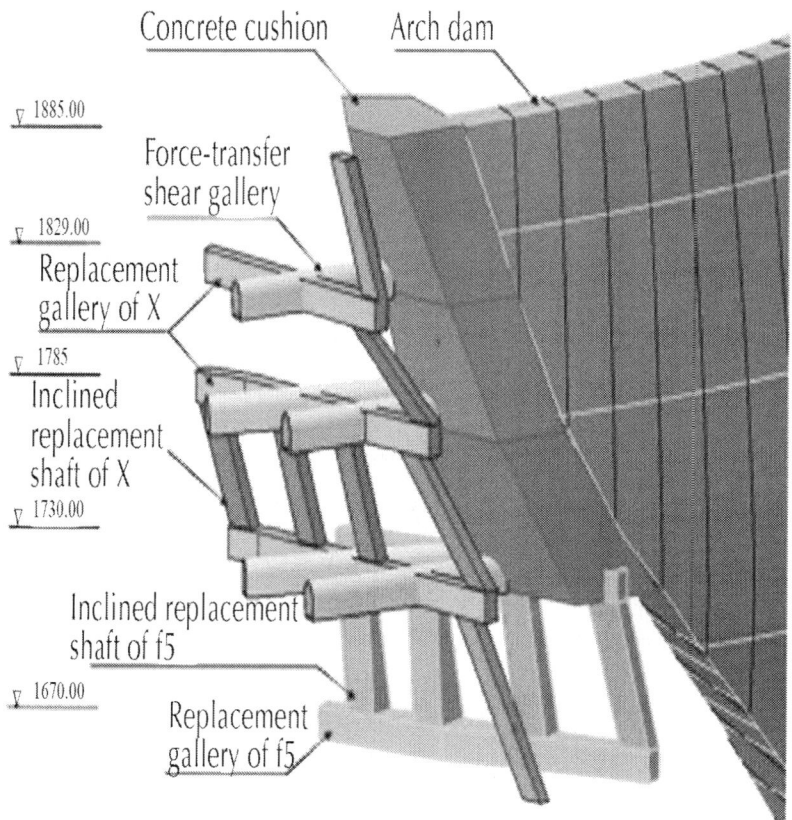

Figure 7: Three-dimensional view of the concrete cushion, replacement galleries and shafts system in left abutment (unit: m).

Fig. 7 shows a three-dimensional (3D) view of the designed strengthening system. It is composed of the concrete cushion, force-transfer shear galleries, and concrete replacement grids desired for the fault f5 and lamprophyre dyke X.

In the right abutment, fault f13 outcrops at the elevations of 1870–1885 m, with a relatively small area, which will be backfilled with concrete after open excavation. In the elevations of 1885–1601 m, inclined anti-seepage concrete replacement shafts will be constructed along the fault plane in consideration of anti-seepage curtains of the dam, with a width of 5 m (Fig. 8).

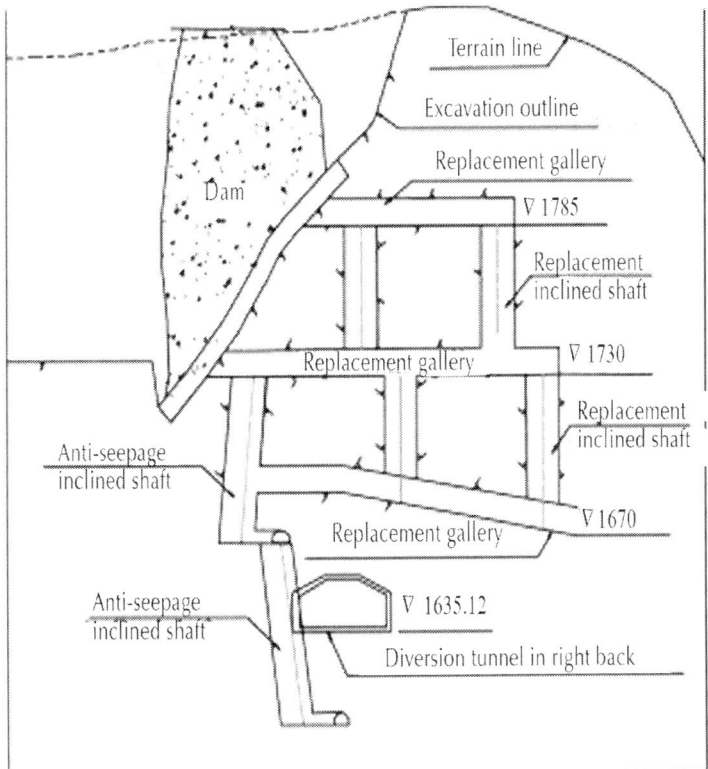

Figure 8: Schematic diagram of the concrete replacement galleries and shafts for fault f14 in the right abutment (unit: m).

As for the fault f14, it outcrops at the elevations of 1740–1790 m, thus it will also be replaced with concrete after open excavation. Regarding of deep-seated parts of the fault, concrete replacement grids and intensive consolidation grouting will be employed, including three concrete replacement galleries at the elevations of 1670 m, 1730 m and 1785 m, respectively, with sectional size of 6.5 m × 10 m (width × height), and five inclined concrete replacement shafts (three between the elevations of 1670 m and 1730 m and two between 1730 m and 1785 m) with a width of 5 m. Intensive consolidation grouting will be implemented in rocks between the elevations of 1670 m and 1730 m, and an inclined concrete replacement shaft will be constructed in the elevations of 1601 m and 1730 m, with a width of 5 m. Thus they can function together with the anti-seepage curtain of the dam (Fig. 8).

Treatment Design of the Slot-Cutting Replacement and Grouting for the Faults f2 and f18 in Lower Part of Dam Foundation and Riverbed

Due to the similar conditions, the treatment measures for the faults f2 and f18 are almost the same as mentioned before. The outcropping parts of the faults will be excavated in consideration of slot-cutting and then be backfilled with concrete. Then cement grouting will be conducted in the foundation area around the deep-seated parts. The design scheme of slot-cutting and excavation with an L-shape for the fault f2, as a typical case, is shown in Fig. 9. Moreover, the fault f2 imposes remarkable adverse effects on the seepage stability of the dam due to the weak and broken rocks in the crushed interlayer zones, a total thickness of 10 m. Additional measures are adopted for the site-specific fault f2, including high-pressure water flushing and grouting after drilling along the crushed interlayer surfaces, and cement-chemical composite reinforcing grouting in the vicinity of the dam and anti-seepage curtains.

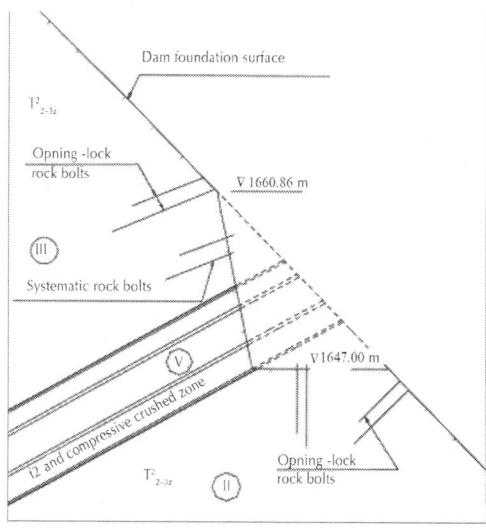

Figure 9: Schematic diagram of the slot-cutting replacement for the fault f2 and the crushed interlayer zones.

Treatment Design of Consolidation Grouting for Weak Rock Masses of the Resistance Blocks in Left Dam Foundation

The resistance blocks in the left dam foundation basically consist of intensively unloaded rock masses grades III_2 and IV. Thus, effective treatment with consolidation grouting needs to be considered for the purpose of stability.

The elevations of grouting range from 1635 m to 1885 m. The lower boundary of grouting extends from weak rock to relatively high-quality rocks below the fault f2, including its interlayer crushed zones. The horizontal boundary is 5–10 m inside the influential zones governed by lamprophyre dyke X. The distance of grouting along the river bank is determined through sensitivity analysis of deformation by FEM.

The sensitivity analysis is conducted by 2D FEM at two typical elevations, i.e. 1750 m and 1830 m. In the calculation model, the concrete cushion, the faults f5, f8, and the lamprophyre dyke X are considered, including the distribution of rocks of various grades. The results indicate that the displacement rate of the arch abutment decreases sharply at the elevation of 1750 m when the grouting distance along the downstream river bank is two times the thickness of the arch abutment (larger than 110 m). The displacements in the upstream and downstream sections are 27 mm and 46 mm, respectively. At the elevation of 1830 m, the displacements in the upstream and downstream sections are 17 mm and 21 mm, respectively, when grouting distance is 2.5 times the thickness of the arch abutment (larger than 95 m).

The results suggest that the distance of grouting along the downstream river bank could be determined in three elevations by considering the thickness of the arch abutment. Above the elevation of 1820 m, the rocks are mainly composed of sandy slate grades III_2 and IV, with considerably low rock quality. Therefore, the grouting range should be considered to somewhat large extent, and the ratio of the distance to the thickness of the arch abutment is determined as 3. In the elevations of 1820–1670 m, the rocks are predominately of grades III_1 and III_2, thus the ratio of 2.5–2.7 can be adopted (Fig. 10). In the elevations of 1670–1635 m, most rocks are of grades II and III_1, and the ratio is determined as 1.5.

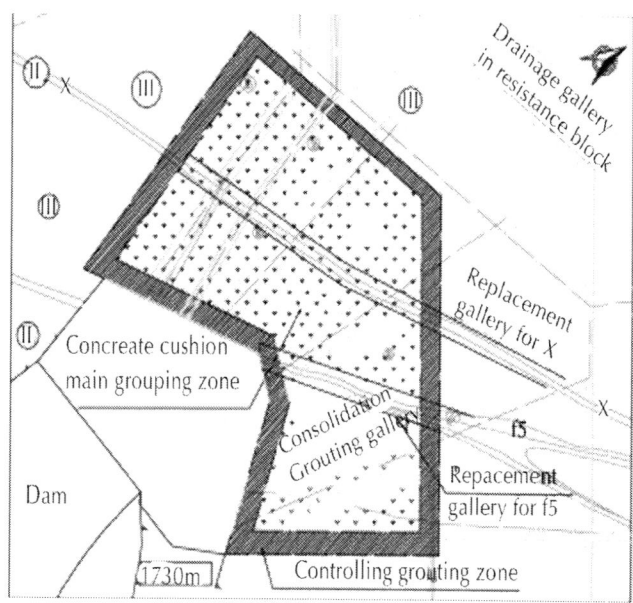

Figure 10: Layout of the grouting area in resistance blocks at the elevation 1730 m.

According to the grouting principle of sliding-resistance, the quality of rocks after grouting should be grades III_2–III_1 or higher. The detailed controlling standards of grouting are listed in Table 4.

Table 4: Requirements of rock quality after grouting

Rocks	V_p (m/s)	E (GPa)	q (Lu)
Marble (III_2)	≥5000	≥7.0	≤3.0
Sandy slate (III_2)	≥4800	≥6.0	
Marble (IV_2)	≥4600	≥5.0	
Sandy slate (IV_2)	≥4500	≥4.2	

Note: V_p is the acoustic P-wave velocity (with guaranteed probability of 85%), E is the deformation modulus of drill cores, and q is the unit permeability rate of rocks.

In order to control the grouting quality and reduce the adverse influence of the grouting on the stability of the abutment slopes, the grouting area in the resistance blocks is divided into two zones, i.e. controlling grouting zone and main grouting zone. The controlling grouting zone is located in the margin of the grouting area with a width of 7.5–10 m. In this region, grouting will be implemented with relatively low-pressure and thick grouting, aiming at forming a protective closure against the main grouting zone. In the main grouting zone, grouting is conducted with high-pressure serum with various water-cement ratios. In particular, controllable grouting with cement mortar is adopted in deep fracture zones which contain large-scale and long fractures and may consume a considerably great amount of grout.

PRELIMINARY EVALUATIONS OF DAM FOUNDATION TREATMENTS

The design schemes of treatments of the dam foundation were proposed in 2009. At present, the arch dam is constructed up to the elevation of about 1830 m. Concrete cushion and other measures for geological defects in the left abutment are almost completed except for grouting at the upper elevations. The impoundment of diversion tunnels has been realized. Results of geo-mechanical model tests, deformation monitoring, acoustic wave examinations of structures after impoundment indicate that treatments of the dam foundation are basically successful, and effectiveness of the design schemes has been preliminarily verified (Varga, 1979, Ronzhin et al., 1986, Xue and Huang, 2005, Mou et al., 2009, Hu et al., 2010 and Song et al., 2011).

Geo-Mechanical Model Tests

Three geo-mechanical model tests for the stability of arch dam and its foundation were conducted in different stages (feasibility study, bidding, and design) before construction (Liu et al., 2003a, Liu et al., 2003b, Zhang et al., 2005 and Jiang et al., 2009). In the tests, the displacement, deformation and cracks of the dam foundation before and after treatment were measured, and the global factor of safety was calculated. Test results show that after treatment, the displacement of the crown was decreased from 105.3 mm to 85 mm under the normal load case. The factor of safety of crack initiation, $K1$, of the dam is increased from 1.5–2.0 to 2.5; the ultimate overload coefficient, $K3$, is increased from 5–6 to 7.5. They are larger than those in most similar projects, which show that the asymmetrically deformable dam has been significantly improved.

Quality Assurance Tests

Quality inspections of grouting in resistance blocks below the elevation of 1730 m were conducted by acoustic P-wave tests, deformation modulus tests in drilling holes, and packer permeability tests. Test results indicate that, after conventional and intensive grouting, the average P-wave velocity in marble of grade III_2 is 5208 m/s, increased by 3.1%; the average deformation modulus is 12.3 GPa, increased by 28.2%. For marble of grade IV_2, the average P-wave velocity is increased to 5184 m/s by 9.1%, and the average modulus to 10.5 GPa by 42.8%. Permeability rates in various rocks are kept in allowable values. It is indicated that all the indexes can meet the design requirements.

Monitoring Results of Deformation, Stresses and Fractures

At present, the reservoir water level remains at about 1702 m. Monitoring results of deformation, stresses and fractures in foundation rocks indicate that the deformations in the dam foundation of both banks and galleries are considerably small with reasonable fluctuation. Distribution of compressive deformation in the riverbed is basically uniform, with a maximum value of 10.3 mm. The apertures of joints between the dam and the foundation are smaller than 1.0 mm. Stresses in the concrete cushion and the dams are mainly of compressive stresses. Monitoring results show that the deformation and stresses in the dam foundation are basically controlled.

CONCLUSIONS

Jinping I arch dam concerns many engineering issues and has attracted worldwide attentions due to its huge scale and unique settings. The geological conditions in the project region are extremely complex, characterized by serious geological defects, which pose great technical challenges on the project stability.

At present, the treatments have almost been completed, and the arch dam is still in the process of construction, and the final impoundment of the reservoir is not experienced yet. The performances of the dam and its foundation are pretty good at present. Numerical analysis, model tests, field excavation, quality inspection, and deformation monitoring indicate that the treatment schemes for geological defects in the dam foundation are effective. The concepts of treatment methods are helpful for similar dam foundations. It is strongly suggested that monitoring of the performances of the dam foundation be continuously conducted. Real-time analysis and dynamic design are needed in order to ensure the safety of the dam.

REFERENCES

1.   Fan et al., 2012, Q. Fan, S. Zhou, B. Li Key technologies of rock engineering for construction of Xiluodu super high arch dam Chinese Journal of Rock Mechanics and Engineering, 31 (10) (2012), pp. 1998–2015 [in Chinese]

2.   Hu et al., 2010, Z. Hu, J. Zhang, Z. Zhou, H. Rao Analysis of stress and deformation of Jinping I high arch dam after foundation reinforcement Rock and Soil Mechanics, 31 (9) (2010), pp. 2861–2868 [in Chinese]

3.   Jiang et al., 2009, X. Jiang, J. Chen, S. Sun Research on physical model test for high arch dam of Jinping I hydropower station Yangtze River, 40 (19) (2009), pp. 76–105

4.   Liu et al., 2003a, J. Liu, X. Feng, X. Ding, J. Zhang, D. Yue Stability assessment of the Three-Gorges dam foundation, China, using physical and numerical modeling-Part I: physical model tests International Journal of Rock Mechanics and Mining Sciences, 40 (5) (2003), pp. 609–631

5. Liu et al., 2003b, J. Liu, X. Feng, X. Ding Stability assessment of the Three-Gorges dam foundation, China, using physical and numerical modeling-Part II: numerical modeling International Journal of Rock Mechanics and Mining Sciences, 40 (5) (2003), pp. 633–652

6. Liu et al., 2004, R. Liu, F. Liu, X. Yu Design of consolidation grouting and treatment of geological defects for the foundation of the Three Gorges dam Water Power, 30 (3) (2004), pp. 18–21 [in Chinese]

7. Liu et al., 2010, M. Liu, R. Huang, M. Yan, F. Lin, J. Huo Preliminary evaluation of geological defects for rock foundation below left bank pedestal of arc dam for Jinping I power-station Journal of Engineering Geology, 18 (6) (2010), pp. 933–939 [in Chinese]

8. Mou et al., 2009, G. Mou, G. Chen, R. Liu Study on Jinping I arch dam left abutment foundation treatment Design of Hydroelectric Power Station, 25 (2) (2009), pp. 7–22 [in Chinese]

9. Qi et al., 2004, S. Qi, F. Wu, F. Yan, H. Lan Mechanism of deep cracks in the left bank slope of Jinping first stage hydropower station Engineering Geology, 73 (1–2) (2004), pp. 129–144

10. Ronzhin et al., 1986, I.S. Ronzhin, L.E. Kanygin, V.N. Chernenko, P.P. Listrovio, F.G. Kim Grout curtain in the foundation of the Nurek dam and evaluation of its effectiveness [Gidrotekhnicheskoe Stroitel'stvo, Trans.] Hydrotechnical Construction, 9 (1986), pp. 18–21 [UDC 624.138.232.1:627.826.3]

11. Song et al., 2010, S. Song, B. Xiang, J. Yang, X. Feng Stability analysis and reinforcement design of high and steep slopes with complex geology in abutment of Jinping I hydropower station Chinese Journal of Rock Mechanics and Engineering, 29 (3) (2010), pp. 442–458 [in Chinese]

12. Song et al., 2011, S. Song, D. Cai, X. Feng, X. Chen, D. Wang Safety monitoring and stability analysis of left abutment slope of Jinping I hydropower station Journal of Rock Mechanics and Geotechnical Engineering, 3 (2) (2011), pp. 117–130

13. Varga, 1979, A.A. Varga The main problems of evaluation of geological structure in dam construction Bulletin of the International Association of Engineering Geology, 20 (1979), pp. 21–23

14. Wang et al., 2012, J. Wang, S. Duan, S. Hu Treatment of high and steep slopes with complicated geological conditions at left abutment of Jinping I hydropower station Chinese Journal of Rock Mechanics and Engineering, 31 (8) (2012), pp. 1598–1605 [in Chinese]

15. Xiang and Rao, 2008, B. Xiang, H. RaoJinping I hydropower station dam site left abutment high slope complicated rock mechanics problems and engineering treatment measures Design of Hydroelectric Power Station, 24 (2) (2008), pp. 14–19 [in Chinese]

16. Xue and Huang, 2005, L. Xue, Z. Huang Finite element method analysis of Jinping I arch dam foundation treatments effects Design of Hydroelectric Power Station, 21 (4) (2005), pp. 11–15 [in Chinese]

17. Zhang et al., 2005, L. Zhang, W. Fei, G. Li, J. Chen, C. Hu Experimental study on global geomechanical model for stability analysis of high arch dam foundation and abutment Chinese Journal of Rock Mechanics and Engineering, 24 (19) (2005), pp. 3465–3469 [in Chinese]

Chapter 10

Moisture Monitoring in Clay Embankments Using Electrical Resistivity Tomography

D.A. Gunn[a], J.E. Chambers[a], S. Uhlemann[a, c], P.B. Wilkinson[a], P.I. Meldrum[a], T.A. Dijkstra[a], E. Haslam[a], M. Kirkham[a], J. Wragg[a], S. Holyoake[a], P.N. Hughes[d], R. Hen-Jones[b], and S. Glendinning[b]

[a]British Geological Survey, Nottingham NG12 5GG, UK

[b]School of Civil Engineering and Geosciences, Newcastle University, Newcastle upon Tyne NE1 7RU, UK

[c]ETH Zurich, Institute of Geophysics, Sonneggstrasse 5, 8092 Zurich, Switzerland

[d]School of Engineering and Computer Sciences, Durham University, DH1 3LE, UK

ABSTRACT

Systems and methods are described for monitoring temporal and spatial moisture content changes in clay embankments using electrical resistivity tomography (ERT) imaging. The methodology is based upon development of a robust relationship between fill resistivity and moisture content and its use in the transformation of resistivity image differences in terms of relative moisture content changes. Moisture level and moisture content movement applications are exemplified using two case histories from the UK. The first is the BIONICS embankment, near Newcastle (NE England), which was constructed in 2005 using varying degrees of compaction of a medium plasticity sandy, silty clay derived from the Durham Till. The second is a Victorian embankment south of Nottingham (Central England), constructed in 1897 using end tipping of Late Triassic siltstone and mudstone taken from local cuttings. Climate change forecasts for the UK suggest that transportation earthworks will be subjected to more sustained, higher temperatures and increased intensity of rainfall. Within the context of preventative geotechnical asset maintenance, ERT imaging can provide a monitoring framework to manage moisture movement and identify failure trigger conditions within embankments, thus supporting on demand inspection scheduling and low cost early interventions.

INTRODUCTION

Engineered slopes, embankments, canals, earth dams, sea walls and flood defences are increasingly susceptible to catastrophic failure due to changes in global climatic conditions and land use. The 4th Assessment Report of the Intergovernmental Panel on Climate Change [23] and [24] predicted that mid- to high-latitude regions can expect more extreme events with up to 20% more precipitation, more flash floods, and a rise in sea levels up to 59 cm by the end of the century. The predicted environmental changes will have inevitable consequences for the serviceability and maintenance of our engineered infrastructure, but while the impact is still largely unknown, we require intelligent platforms and science to monitor current condition and assess risk over the whole life cycle of UK assets. Aged assets include: Canal & River Trust/Scottish Canals with 3450 km of aged canal earthworks, Network

Rail with over 20,000 km of earthwork embankments and cuttings, and London Underground with 236 km of embankments and cuttings in Greater London, all contributing significantly to the UK economy.

A significant number of UK earthworks between 100 and 200 years old were constructed using tipping methods, which was standard in the 19th century. This has left a legacy of ageing, highly fissured, weak and heterogeneous earth structures, which are still intensively used but prone to failure under aggressive climatic stresses [28]. Common problems in certain subgrade soil types include shear failure and mud pumping caused by loss of strength and cohesion [33] and [25], heave, deformation and the formation of ballast pockets [31] and [5], which also occur in zones of low density and stiffness. In most cases, subgrade problems are associated with high moisture levels, a key factor in reducing consistency and strength, and ultimately leading to failure [32] and [18].

Modelling undertaken during the recently completed FUTURENET project [3] and [4] showed how climate or weather event sequences affect the traffic capacity of transportation networks. Weather events have direct effects on the permanent way such as increased temperature on risk of track buckling (or pavement rutting) and related effects on potential failure of the subgrade and surrounding ground including landslide, shrink–swell and scour. Climate resilience planning for transportation networks requires access to near real-time, volumetric, and hence holistic, assessment of infrastructure condition, including ground water movement and the moisture levels within the earthworks asset. Maintenance practice, based primarily on surface observations, is a barrier to proactive approaches because these represent the latter stages of failure and reinforce responsive solutions. Risk-based prevention and early interventions require identification of the incremental development of internal conditions that ultimately trigger failure. Key to this process will be adaptive technologies delivering real-time images of the true 3D spatial variation of groundwater and geotechnical properties affecting stability. While providing useful ground truth, a full understanding of vital ground processes with sufficient temporal and spatial resolution is often not possible from invasive investigation alone. We assert that this role can be filled by non-invasive geophysical methods that not only provide real-time images of moisture movement but are also calibrated so as to indicate full 3D, quantitative geotechnical property changes. This can be achieved if the geophysical relationships between electrical resistivity and geotechnical properties (such as moisture content, pore pressure and strength) are well understood.

Resistivity imaging, or electrical resistivity tomography (ERT), is sensitive to lithological and mineralogical heterogeneity [34] and changes in ground temperature and soil moisture content [10], [11], [19] and [12]. In locations where lithology and mineralogy are unchanged, provided ground temperature effects can be corrected, changes in successive ERT surveys over an electrode array of constant geometry and location will be due to ground water movement and subsequent moisture content variations. Thus, by applying appropriate temperature correction and petrophysical relationships linking resistivity and saturation [7],[6] and [12], time-lapse, volumetric (4D) images of water movement and moisture content changes can be constructed from repeated ERT surveys. Alongside the increased use of ERT in site investigation, purpose built ERT monitoring instrumentation has rapidly developed and now incorporates telemetric control and automatic data transfer, scheduling, and processing [30]. This type of instrumentation is now being applied to monitor of natural slopes [27], [37] and [35] and transportation earthworks [19] and [12].

In this study we describe repeat survey-based approaches using standard field equipment/return visits and fully automated monitoring and data capture on permanent field installations to investigate the structure and processes in sections of two embankments. We provide two case histories: firstly, from the BIONICS research embankment, Nafferton Farm, Northumberland, UK [14] and [21] constructed using varying amounts of compaction in 2005 from sandy, silty clay derived from partially sorted Durham Till; which includes identification of individual lifts from 2D resistivity sections across the embankment transect; and secondly, from an embankment along the former Great Central Railway near East Leake, Nottingham, UK[2], [17] and [19] constructed via end-tipping of materials derived from the East Leake Tunnel cutting to the south; which includes identification of fill regime changes in a 2D resistivity section along the axis of the embankment, dynamic, seasonal wetting and drying fronts moving through a 2D transect of the embankment and a demonstration of the potential application of 3D volumetric images of moisture movement and geotechnical property visualisation for planning maintenance. Finally, these case histories provide the context for a broad discussion relating to the foundation for new risk-based asset management practices incorporating automated, electrical imaging technologies into early intervention decision processes, such as proactive drainage planning.

SOIL AND ROCK RESISTIVITY

Resistivity Measurements and Field Systems

Fig. 1a shows that the resistivity, ρ_s of a unit volume of material is given as,

$$\rho_S = \frac{V}{I} \cdot \frac{A}{L} \tag{1}$$

where $\frac{V}{I}$ is the ratio of the difference, V in the electrical potential at the two opposing faces of a unit cube that are orthogonal to the

current flow, I and is equivalent to the material resistance, R and $\frac{A}{L}$ is the Geometric Factor (in Fig. 1a) that accounts for how the current flow within the material and the measurement are affected by the electrode geometry, and converts resistance R to resistivity, ρ_s.

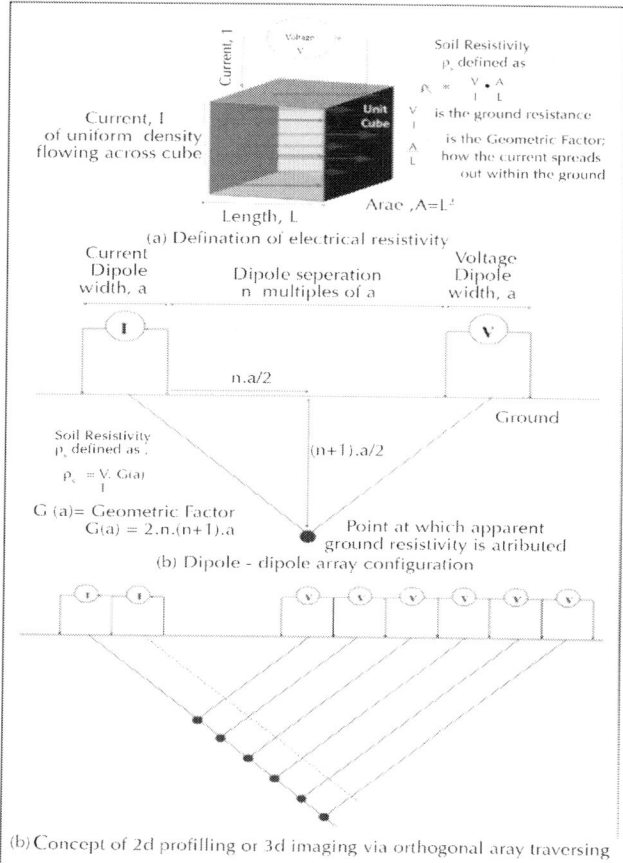

(a) Defination of electrical resistivity

(b) Dipole - dipole array configuration

(b) Concept of 2d profilling or 3d imaging via orthogonal aray traversing

Figure 1: Resistivity, field surveying arrays and constant separation traversing methods.

Resistivity is measured in the field using a four-electrode array consisting of two current injection electrodes and two potential measurement electrodes. In general, the depth of investigation increases with increasing electrode separation, where the different electrode array configurations determine the specific relationship. For example, Fig. 1b shows how the depth of investigation for a dipole–dipole array is related to the common spacing (denoted 'a') between the current and potential electrode pairs. It also shows how a 2D 'apparent resistivity' section along a transect can be constructed from a series of resistivity measurements at different inter-dipole spacings (denoted 'n'). Further processing can also be undertaken to refine these images to produce the best estimate of the true ground resistivity distribution; a process termed 'inversion'. The ABEM SAS 1000 is typical of the field equipment used to make resistivity-depth soundings or 2D cross-sectional surveys. A series of field measurements are made, usually by varying the electrode spacing in standard four-electrode array configurations, such as the dipole–dipole (or Wenner or Schlumberger) arrays, from which apparent resistivity sections are constructed. The voltage measurement between two potential electrodes can be considered as a single channel. As surveys require multiple measurements, the duration of the survey can be reduced by an equivalent factor to the number of channels used.

Datasets for 3D imaging typically require many thousands of four-electrode measurements over a range of geometries to be carried out across the area of interest. Thus, equipment with lower numbers of input channels are disadvantaged by longer survey times. The AGI SuperSting R8 is typical of field equipment used for 3D surveys, boasting eight channels with the potential to connect up to 65,000 electrodes, (although most surveys don't utilise anywhere near this potential but the eight channel system reduces surveying times). 3D apparent resistivity images, or models of the true resistivity distribution in the subsurface are constructed from the measured resistivity dataset, in a similar manner to the multi-point construction in 2D surveys (Fig. 1c). The new generation of remote monitoring platforms such as the Automated time-Lapse Electrical Resistivity Tomography (ALERT) and the very recent Proactive Infrastructure Monitoring and Evaluation (PRIME) systems combine emerging electrical resistivity imaging technology with wireless telecommunications, server-based processing, site databases and web portal access [29] and [38]. These platforms provide the basis for "smart" technology capable of monitoring the internal physical condition of embankments using diagnostic imaging methods. They operate in the same manner as the field resistivity instruments but are remotely controlled via wireless telecommunications, such as over the mobile phone network. They provide the potential for high-resolution images of subsurface structure, and when used in time-lapse mode, these platforms can monitor groundwater movement and changes in the moisture content of earthworks and surface movement in near real time [37] and [12]. Hence, these platforms capture information about subsurface processes (groundwater movement) and the resulting spatial and temporal changes in subsurface geotechnical properties, such as moisture content. If we denote these phenomena as the 'CAUSE', we can now access (visualise) information relating to the 'CAUSE' in synchronous with remote, high resolution measurements of surface movement, which we shall denote the 'EFFECT' (or similarly we could apply the term 'SYMPTOM'). Remote access delivers a 'virtual earthworks asset', where the delivery of this information into maintenance decisions, and how it is used to support early interventions, will be core to the development of true preventative maintenance practices (which we develop further in Section 6).

Resistivity–Moisture Content Relationships

General resistivity ranges for commonly occurring rocks and soils are presented in Fig. 2. While ground resistivity is dependent on the composite soil or rock, it is also controlled by the amount of moisture stored within the pore space and the ionic distribution about grain surfaces (hence the geological materials plotting across a range of resistivities in Fig. 2).

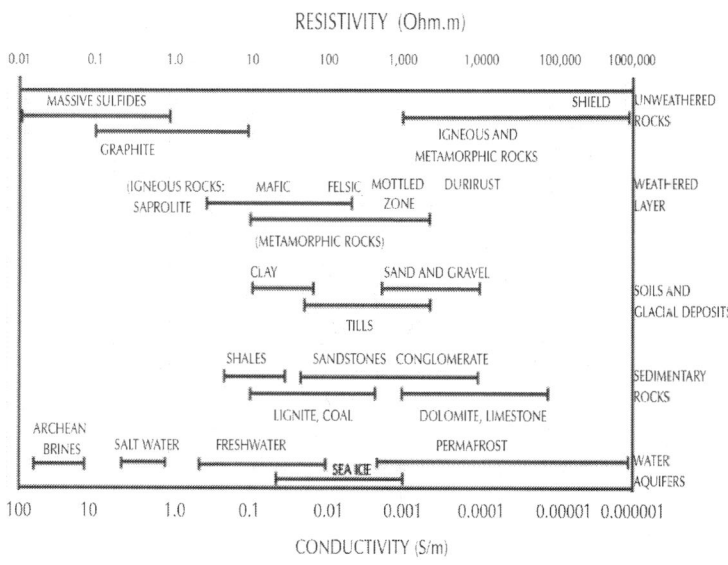

Figure 2: Resistivity ranges for surface waters, rocks and soils.

In sands and gravels, the current flows around non-conducting grains via ionic migration within the saturating fluid. A clear relationship has been established between resistivity in sands and gravels and various other factors (Granular: in Fig. 3a), so an accurate measure of resistivity can lead to the calculation of key soil parameters, particularly pore water saturation, and therefore moisture content. This relationship is often termed Archie's Equation [1], where the soil resistivity $_s$ is related to the resistivity of the fluid in the pore space, $_w$ by the degree of saturation, S, i.e. the proportion of the pore space that is filled by the fluid, where $S = 0$ represents completely dry soil (air filled pores) and $S = 1$ represents fully saturated soil (fluid filled pores). Archie's equation also shows that soil resistivity increases with greater compaction (via the compaction factor, 'a' in Fig. 3a), but decreases with increased porosity, n, where the value of the exponent m is related to the grain morphology and how it affects current flow.

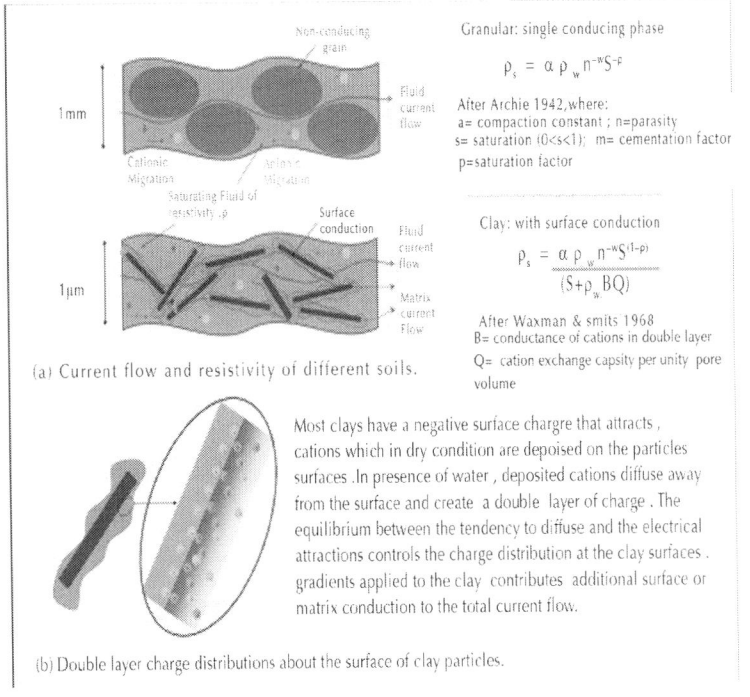

(a) Current flow and resistivity of different soils.

Granular: single conducing phase

$$\rho_s = \alpha \, \rho_w \, n^{-m} S^{-p}$$

After Archie 1942, where:
a= compaction constant ; n=parasity
s= saturation (0<s<1); m= cementation factor
p=saturation factor

Clay: with surface conduction

$$\rho_s = \alpha \, \rho_w \, n^{-m} S^{(1-p)}$$
$$(S + \rho_w BQ)$$

After Waxman & smits 1968
B= conductance of cations in double layer
Q= cation exchange capsity per unity pore volume

Most clays have a negative surface chargre that attracts, cations which in dry condition are depoised on the particles surfaces .In presence of water , deposited cations diffuse away from the surface and create a double layer of charge . The equilibrium between the tendency to diffuse and the electrical attractions controls the charge distribution at the clay surfaces . gradients applied to the clay contributes additional surface or matrix conduction to the total current flow.

(b) Double layer charge distributions about the surface of clay particles.

Figure 3: Moisture content, its effect upon soil charge distribution and resistivity.

Generally, clay resistivity is far lower than granular (e.g. sand and gravel) soil resistivity due to additional matrix conduction caused by the movement of ions distributed across the surfaces of clay particles (clay: in Fig. 3a). Clay resistivity is controlled by both mineralogy and cationic exchange capacity and can also be related to moisture content using established relationships (in Fig. 3a). This relationship was developed by Waxman and Smits [36], where the numerator relates to the ionic migration in the saturating fluid and has a similar form to Archie's Equation, but where the denominator relates to the conduction contribution through the clay matrix. The B parameter relates to the conductance of the cations (such as potassium, calcium, sodium or aluminium) and Q relates to the exchange capacity (CEC) or the capacity for the clay to hold cations within the diffuse double layer about the clay surface (Fig. 3b). For example, higher resistivity clays, such as kaolinite have a low CEC, lower resistivity clays, such as chlorite and illite have a medium CEC and the lowest resistivity clays like smectite have a high CEC. As the plasticity index generally increases with increased CEC, the resistivity ranges for clay dominated mudstones provide not only a very useful index for moisture content but also a very useful proxy for shear strength and thus, resistivity imaging carries the potential to be used to monitor ground strength and stability.

Resistivity has become an important engineering property because it can be used to derive the volumetric moisture content in the calculation of soil moisture deficit (SMD), a standard index of groundwater saturation used in the transportation industry. The railway industry currently uses a simplified calculation of SMD based on regional rainfall using the Meteorological Office Rainfall and Evapotranspiration Calculation System (MORECS). This method provides a broad classification of the network based on a km grid scale and takes no account of either the proportion of precipitation entering the groundwater system or its actual subsurface movement or distribution. However, new geoelectrical imaging-based technologies can provide dynamic 3D images of SMD based on real-time monitoring of the actual moisture movement within infrastructure. Also, resistivity imaging can be used to map the spatial and temporal changes in moisture content, enabling real-time assessment of plasticity changes, for example in response to sustained drought or rainfall.

STUDY SITES

Electrical resistivity remote monitoring systems have been installed at two earthworks embankments. Field resistivity data were collected at both sites using the dipole–dipole array configuration (Fig. 1), and apparent resistivity images were inverted using Res2DInv or Res3DInv software [26].

East Leake Site

The East Leake research site comprises a 140 m long section of the whole embankment on the former Great Central Railway (GCR) that extends 800 m. The embankment was built up over the Branscombe Formation of the Mercia Mudstone Group in 1897 using local materials excavated from cuttings to the SW and NE. The material was tipped and then compacted by subsequent movement of shunting locomotives and tipping wagons across the tipped material. The tipping method used along this section of the line was not stated explicitly by Bidder [2], but has been deduced to have been end tipped (e.g. based on historical photographs taken by S.W. Newton of end tipping wagons and operations in the vicinity of the site at the time of construction). The embankment has been subject to several phases of site investigation spanning from September 2005, which has included drilling, collection of core samples, invasive probing and non-invasive geophysical surveying [16], [17], [18], [19], [8], [9], [10] and [11]; borehole locations and the resistivity lines are shown Fig. 4. These phases of SI have shown the embankment to be highly heterogeneous and Gunn et al. [19] provided an interpretation of an along axis section through the test site based upon an interpretation of pits, borehole logs and small strain stiffness profiles, derived from surface wave surveys.

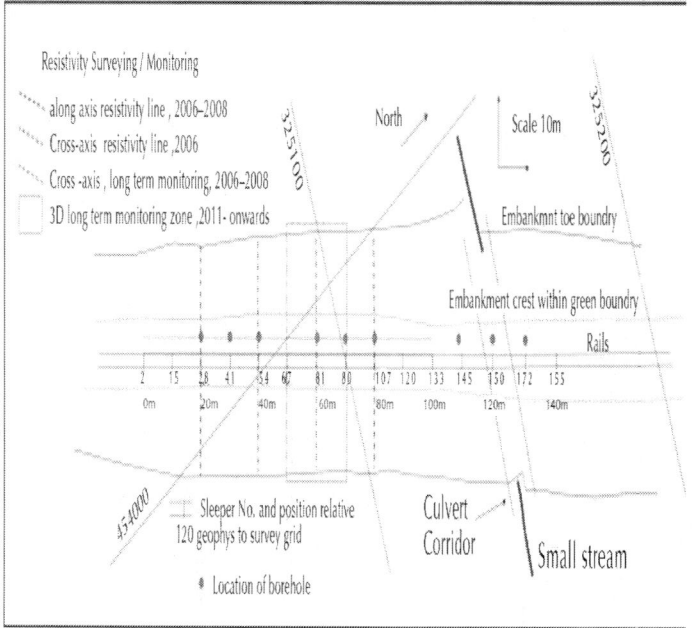

Figure 4: East Leake: layout of resistivity surveying and monitoring lines on a section of embankment.

Across the site, soiled modern ballast generally occurs from the surface to around 0.5 m. Immediately underlying the modern ballast in the SE half is the original engineered ballast pavement as described by Bidder [2] and Fox [13] comprising angular granodiorite gravel over granodiorite cobbles. Glaciofluvial sand and gravel occurs beneath the modern ballast over the NE half. The sand is generally uncemented but occasionally the sand was bound within layers around 100 mm thick by fine, white, powdery non-carbonate cement believed to be gypsum leached from other fill materials. Siltstone appears to have been used as an original final dressing to the earthworks fill prior to the laying of the original ballast, but has degraded *in situ* in the embankment. It occurs across much of the section apart from the furthest 30 m in NE end where it is believed to pinch out into the glaciofluvial sand and gravel. All of these materials overlie degraded Late Triassic mudstones that make up the bulk of the earthworks fill either comprising dark grey-black Westbury Mudstone and Clay or red-brown Branscombe Mudstone and Clay.

At East Leake, installations included an array of 64-electrodes spaced at 1.5 m that ran parallel to the west rail ('along-axis' black dashed line in Fig. 4). The electrodes were inserted into a shallow trench that ran along the crest, offset from the rail by approximately 2.5 m, which was excavated with a narrow bucket to approx. 300 mm deep and covered over with ballast. Also, several 32-electrode line arrays were installed across the embankment, each spanning from the toe of the west flank to the toe of the east flank with a 1 m spacing ('cross-axis' blue and red dashed lines in Fig. 4). Along the earthworks flanks, electrodes were installed into a slit cut with a small spade. The positions of these cross-sectional transects were chosen to investigate the effect of different fill materials on the resistivity sections. These lines were installed during July 2006, when a series of resistivity measurements were first made using an AGI Super Sting R8 system. From that time for a period extending into early 2008, repeat measurements were made on the along-axis line and the cross-axis line at the 60 m station at approximately 6 weekly intervals using the same equipment. Measurements were made using the dipole–dipole configuration with current and potential dipole widths (a) of 1–4 times the electrode spacing and unit dipole separations (n) of 1–8.

During September 2010, a permanent resistivity monitoring array was installed within a 22 m section of the embankment, comprising twelve cross-axis lines spaced at 2 m intervals. This 3D zone was approximately centred on the existing cross-axis line at the 60 m station (red dashed line in Fig. 4). The additional lines also comprised 32 electrodes spaced at 1 m intervals, running from the toe of the eastern flank to the toe of the western flank. An ALERT system was also installed at the site at this time, with which measurement schedules were variably programmed to capture (with high temporal resolution) weather event triggered water movement within the embankment. In this way, response to heavy rainfall events could be studied. The system was powered by a combination of solar panel and a methanol fuel cell charging banks of 12 V batteries. Remote monitoring over this 3D array was undertaken over the period from September 2010 to February 2012. A network of proprietary temperature sensors was also distributed in the embankment to depths of 3.7 m, which are used to study the seasonal temperature change patterns throughout the embankment.

BIONICS SITE

The embankment axis is orientated in an east–west direction. It is 90 m long, 6 m high, has a 29 m base width and a 5 m wide crest with 1 in 2 slopes on the flanks. The embankment was constructed in 2005 in four main 18 m-long sections, with the two inner-most sections constructed according to Highways Agency specifications using 0.3 m lifts and 18 passes of a 7.3 tonne self-propelled smooth drum vibrating roller [14],[15], [20], [21] and [22]. These have been termed the 'well compacted panels' and simulate new-build highway embankments (Panels B and C in Fig. 5a). The two outer-most sections were built to represent poorly constructed/heterogeneous rail embankments, using four lifts, each nominally of 1.3 m height with minimum tracking by site plant; termed 'poorly compacted' (Panels A and D in Fig. 5a).

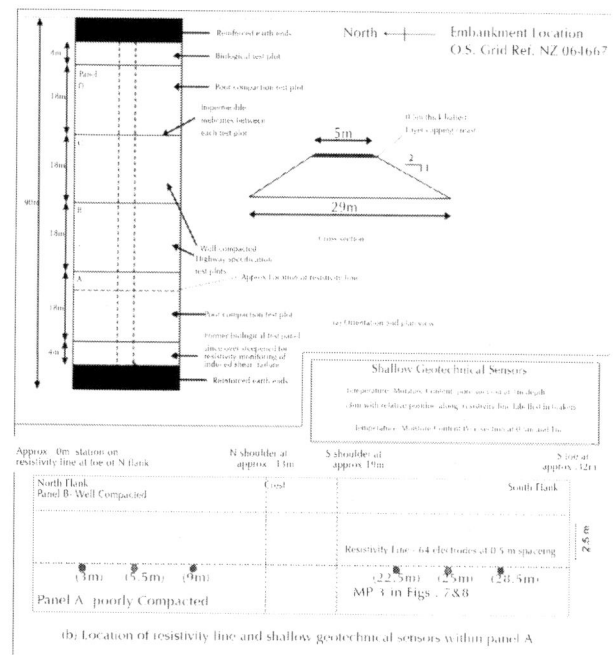

Figure 5: BIONICS: general structure of embankment and layout of resistivity line in poorly compacted section.

To prevent any hydraulic connectivity between each section vertical impermeable membranes were installed during construction between the panels. Immediately after construction the embankment slopes were seeded with grassland seeds typical of the North East of England and other plant species allowed to colonise the embankment naturally. The earthworks fill comprise a sandy, silty clay, which was a locally sourced glacial till (Durham Till) with matrix supported clasts (of greater than coarse gravel size) removed. Atterberg limit tests indicated moisture contents at the Plastic Limit (PL) of 24% w/w (approx. 38% v/v) and Liquid Limit (LL) of 45% w/w (approx. 72% v/v), which classifies the fill material as intermediate plasticity. The dry density of material dried from around the PL is approximately 1.6 Mg/m^3. The crest was capped with a 0.5 m thick layer of basalt ballast.

64 electrodes, spaced at 0.5 m were installed across a 32 m transect from the toe of the north flank to the toe of the south flank (Fig. 5b). In the silty clay earthworks, electrodes were installed into a slit cut with a small spade, while across the crest, electrodes were bedded into bentonite clay that filled 0.3 m deep, fist-sized pits, which were then re-covered with ballast. Proprietary geotechnical sensors including the Decagon 5-TM temperature and moisture content and the MPS-1 water potential (suction) were installed just off-line at depths of 0.5 m and 1 m (below the surface) at three locations on the south and north flanks as shown in Fig. 5b. Installation phases occurred during November 2008 and October 2009 and the period of resistivity measurements on this array extended up to mid-2011. Within this monitoring period, a permanently installed ALERT system was used to make resistivity measurements over this line on a weekly interval. Measurements were made using the dipole–dipole configuration with current and potential dipole widths (a inFig. 1b) of 1–4 times the electrode spacing (0.5, 1, 1.5 and 2 m) and unit dipole separations (n in Fig. 1b) of 1–8 times the electrode spacing.

SURVEYING AND MONITORING IMAGES

Using the aged, end-tipped embankment and the modern compacted embankment as case histories, this section presents examples of how 2D, 3D and time-lapse resistivity difference images can be used to aid interpretation of embankment structure and condition, and monitor ground water movement processes through the earthworks. Most importantly, these case histories demonstrate the potential for using resistivity as a proxy for the long-term monitoring of geotechnical properties, offering insight into future technology that can provide timely information to support preventative asset maintenance practices.

2D Static Images – Material Mapping Application: East Leake Site

The 2D along-axis section provides infill information between boreholes on the subsurface structure to aid interpretation relating to the construction of the embankment (Fig. 6). In Fig. 6a, from the 0 m to the 40 m stations, the resistivity of the interval from 0.8 m to 4 m is generally below 20 Ω m and this is consistent with values that would be expected for clay and mudstone materials. This zone of low resistivity coincides with low stiffness, low penetration resistance zones and relatively high friction ratios, and has been classed as a zone of high moisture and low strength [16], [17] and [19]. The originally tipped fill would have been a coarse gravel comprising lithoclasts predominantly of locally sourced Westbury Mudstone (Fig. 6b). Over the lifetime of the embankment (116 years), the mudstone has weathered to clay and it is believed that this degraded clay material is a key factor in the moisture retention in this zone. From the 40 m to about the 60 m station, the resistivity of the interval from 0.8 m to 4 m is between 20 Ω m and 50 Ω m. This has been interpreted as fill predominantly of gravel comprising Westbury Mudstone, possibly with occasional siltstone from the Blue Anchor Formation. Lower resistivities within this range are consistent with less weathering and less degradation of the mudstone clasts resulting in the earthworks being more freely draining. This zone has been classified as intermediate strength and moisture content, and, represents a buffer between the low strength, mudstone, clay-dominant fill and the high strength, sand, gravel and siltstone-dominated fill. This buffer zone provides the interface between earthworks with very different engineering properties and hence very different performances including response to dynamic loading, drainage, and seasonal variation. From 60 m a lens of fill comprising sand, gravel and siltstone produces a wedge shaped zone with resistivities above 150 Ω m. The wedge develops from the surface at about the 40 m station and thickens to about the 70 m station such that it extends from just beneath the surface to 4 m depth. This high resistivity wedge persists longitudinally under the embankment crest over this depth interval towards the 100 m station. The high resistivity of this zone indicates that the fill has low moisture content and has been shown to be associated with high penetration resistance values and high stiffness values [16], [17] and [19].

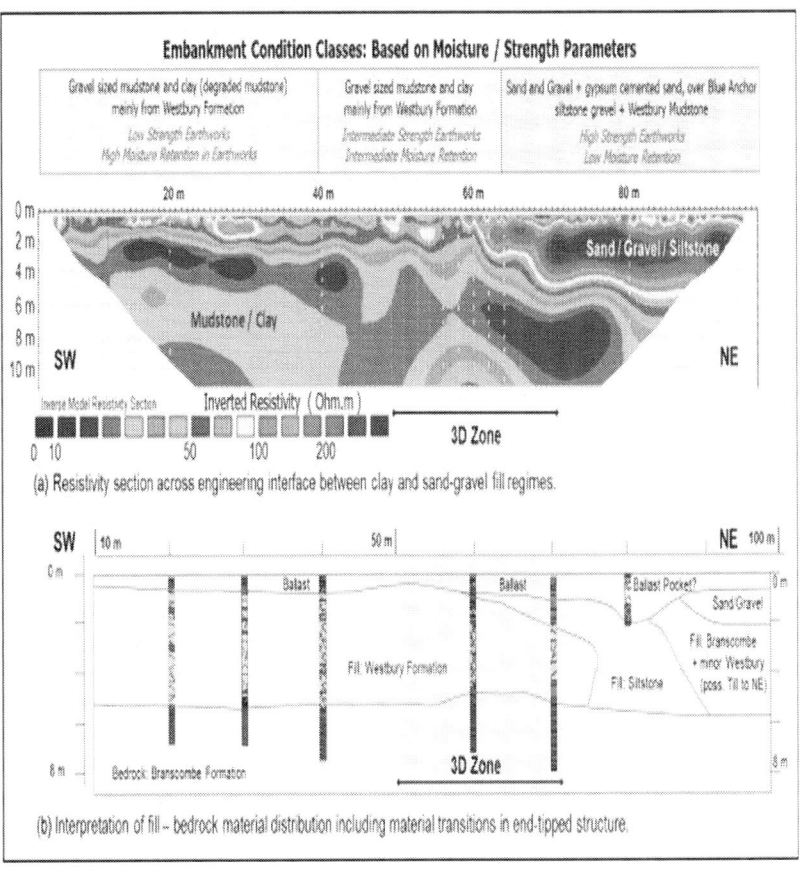

Figure 6: Structural interpretation and material characterisation aided by 2D 'along-axis' resistivity image.

2D Dynamic Difference Images – Groundwater and Moisture Content Changes: East Leake Site

Fig. 7 shows how the record-breaking rainfall during the 2007 winter–summer transition was captured by the 6-weekly schedule of resistivity image monitoring on the 'cross-axis' line at the 60 m station. This extreme rainfall event led to the gradual infiltration and near full saturation of the east flank of the embankment (left side). In fact, the infiltration zone extends into the underlying Branscombe Formation, implicating bedrock processes as part of the drainage problem. During these flood events, standing water develops in this area within the cess at the toe of the east flank. While, the embankment at this location shows little sign of climate-induced distress, the sequence of images in Fig. 7a demonstrates how resistivity-differencing between images can be used to monitor build up of potentially unstable moisture conditions. Chambers et al. [12] describe methods of temperature correction and development of moisture content–resistivity relationship for the Westbury Formation fill materials at the East Leake site. The relationship is based upon fitting the Waxman-Smits [36] relationship (Fig. 3a) to a series of resistivity measurements on dried samples that were reconstituted to a range of known moisture contents. Based on this resistivity-moisture content transformation, Fig. 7b provides an imaged estimate of the saturation distribution throughout the embankment after a heavy rain event on 30th March, 2010. The key features of note include: the infiltration into the east and west flanks (which appears greater on the east flank) and the highly saturated central core of the embankment, which is believed to partly associated with perching over an interval of clay degraded from the Westbury Mudstone. In the context of early intervention, the high levels of saturation may be sufficient to classify this location as at risk, requiring monitoring that could be achieved via remote delivery of saturation images on a weekly or even daily basis from a temporary retrofitted electrode array. The monitoring period would then be sufficient to capture the full extremes of weather events affecting the site. The temporal and spatial characteristics of the groundwater movement would be investigated using a series of time-lapse images and used to plan future drainage schemes.

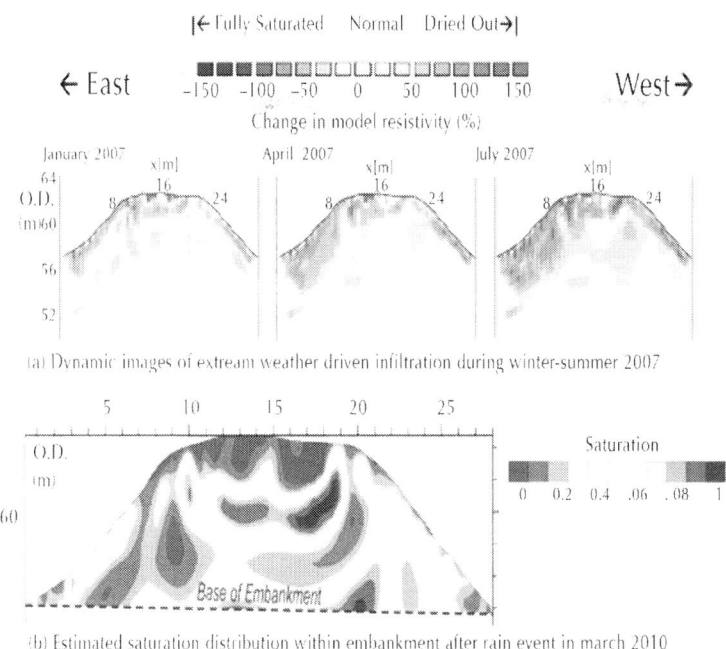

Figure 7: Imaging moisture movement and moisture conditions within embankment.

2D Dynamic Difference Images – Moisture Content and Pore Suction Changes: Bionics Site

Fig. 8 shows a series of six resistivity images across the BIONICS embankment captured from spring to summer of 2009 when air temperatures regularly exceeded 20 °C during May and June. During this time there was little significant rain until a series of weather events that brought rainfall that occurred in June and July. The layered structure in the resistivity sections can be attributed to the sequence of lifts and ballast capping employed during the construction of this poorly compacted section of embankment, where Table 1summaries the broad resistivity and geotechnical properties of the layers identified.

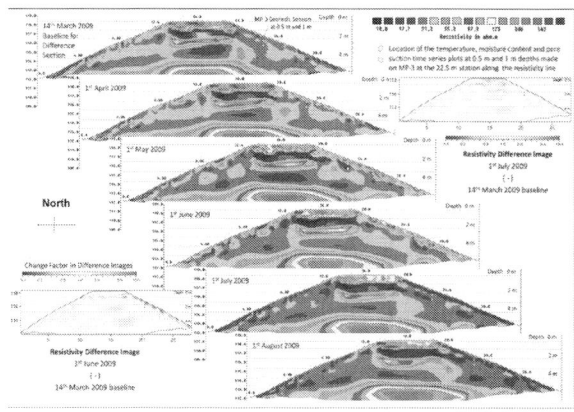

Figure 8: BIONICS: resistivity image differences during drying event of spring–summer 2009.

Table 1: Resistivity and estimated geotechnical property ranges of the BION-ICS embankment layers

Structure			Geophys	Estimated geotechnical property ranges		
Layer no.	Depth interval (m)	Main fill lithology	Resistivity range (Ω m)	Moisture content range (% v/v)	Plasticity	Shear strength (kPa)
1	0–1	Ballast	800–170			
2	1–2	Clay	40–<10	25–80	<PL LL	100–10
3	2–3	Clay	170–50	5–20	≪PL	≫100
4	3–4	Clay	30–20	30–45	PL	⏃100
5	>4	Clay	>30[a]	<30	<PL	>100

PL – Plastic Limit, LL – Liquid Limit.

a Images have lower sensitivity in centre of embankment at depth.

The ballast cap at the top of the embankment (Layer 1) exhibits the largest resistivities to 800 Ω m. The underlying silty clay in Layer 2 exhibits the lowest resistivities of below 10 Ω m. Low resistivity zones occur at the interface between Layer 2 and the overlying ballast, where infilling of the ballast cap into a depression is observed beneath the northern half of the embankment crest. The low resistivities in layer

2 below this structure would result from perching of rainwater that drains through the ballast. The clay fill in this zone would be soft, being above its PL (possibly approaching its LL) and of low strength ≅ 10 kPa. Notably, Layer 3 is highly resistive, indicating a high strength clay with moisture contents well below the PL. The moisture contents in Layer 4 and the underlying zones appear to close to the expected in situ moisture levels and show little variation during this monitoring period.

This series of images capture the property change domains within the embankment associated with it drying out in response to relatively low rainfall and a seasonal increase in temperature from the spring to summer 2009. The resistivity difference images (between the baseline image on 14th March and the 1st May, 1st June) indicate the resistivity change distribution throughout the embankment, and thus provide insight into the exfiltration process. Note that the resistivity of Layer 2 appears to decrease and Layer 3 increase during this period, which could be related to a wicking suction due to evaporation from the ballast cap. However, the most significant increase in resistivity (as high as 10 times) occurred in the upper 0.75 m interval of the south flank. During this drying event, the ground temperature rises from around 6 °C (6 °C) in March to above 14 °C (12 °C) in June at 0.5 m (1 m) depths, (Fig. 9).

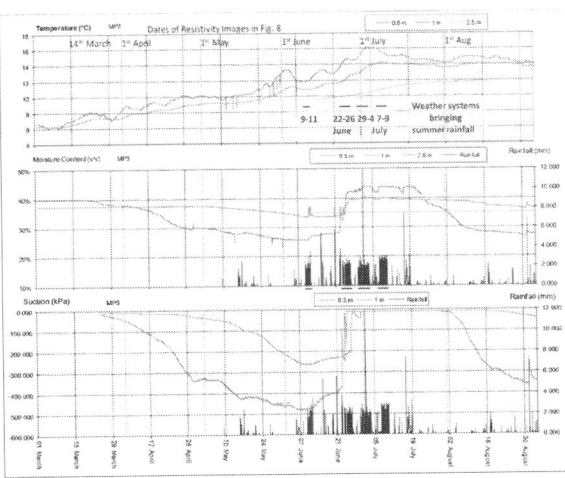

Figure 9: Temperature, moisture content and pore suction time series at MP-3 0.5 and 1 m depths during spring–summer 2009 drying event.

During early March, the moisture content sensor recorded 40%v/v (37%) and the pore suction was at its limit of suction of −10 kPa (−10 kPa) at 0.5 m (1 m) depths. Changes in the moisture content and pore suction measurements during drying throughout late March to early June to 26%v/v (34%) and −480 kPa (−270 kPa) at 0.5 m (1 m) show how the moisture loss and development of suction pervades from the surface into the embankment. Lowest moistures and greatest suctions were recorded on 8th June prior to a series of weather systems that brought steady rainfall, the first of which occured from 9 to 11 June. Recharge into the embankment reverses results in an increase in the measured moisture content and a decrease in suctions. There is a more immediate and greater magnitude of response at 0.5 m than at 1 m, again demonstrating how infiltration is driven by recharge from the surface. Note how the later rainfall events during 22–26 June and 29 June–4 July resulted in a loss of suction and higher moisture contents in July than in March. These events provided sufficient rain to produce lower resistivities across the whole of the embankment in the upper 0.5–1 m interval (resistivity difference image between 1st July and 14th March).

3D Dynamic Images – Process and Property Change Visualisation

The comparison of time series data with dynamically changing 2D or 3D images demonstrates the challenge of up-scaling from a single point sensor to the whole asset. Although it is common practice to monitor earthworks using point sensors, it is very difficult to fully quantify the processes driving property changes (i.e. magnitude and rate of spatial change) without a dynamic, volumetric visualisation of those properties throughout the whole earthworks asset. Full volumetric visualisation can be provided by a 3D image, but while technologies for direct, non-invasive geotechnical property imaging are scarce (or if they exist at all), resistivity-based proxy images to moisture content and even pore suction are possible provided sufficiently robust relationships exist between these properties. 3D resistivity images are very easily realised by electrode arrays over a surface area, such as by using a number of parallel line arrays as was the case at East Leake. The resistivity-moisture content relationship used for interpreting the 2D sections can also be applied to 3D images, such as shown in Fig. 10.

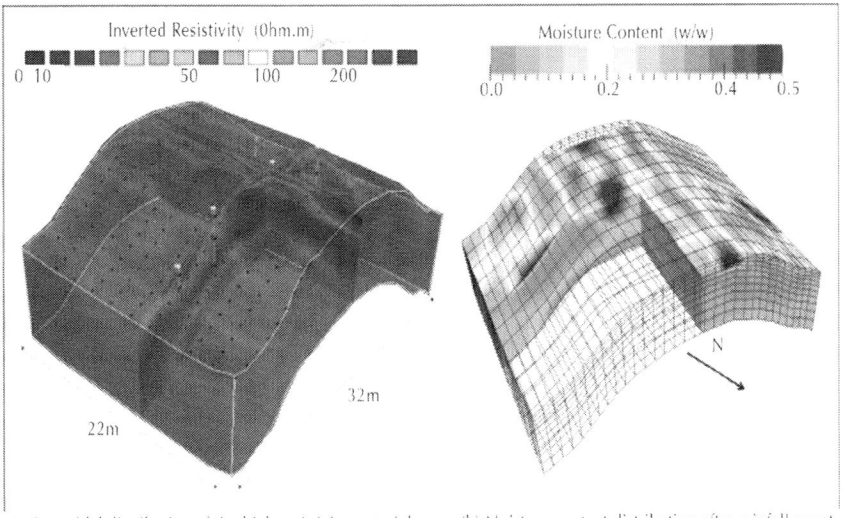

(a) Lensoidal distribution of the high resistivity materials   (b) Moisture content distribution after rainfall event
under the creast and across the flanks of the embankmemnt.   in Februry 2011 showing infiltration into crest

Figure 10: 3D resistivity and derived geotechnical 'proxy' property distributions in aged Victorian earthworks, (after [12]).

The central along-axis 2D section (non-transparent) in Fig. 10a features the proportion of the section shown in Fig. 6 between the intermediate and higher strength zones. Fig. 10b shows the moisture content distribution after a minor rainfall event in February 2011. While, much of the crest is at high moisture content, very high moisture zones can be observed, such as centre crest on the east–west cutaway. Also, shallow zones of very high moisture on the flanks probably indicate gaps in the vegetation cover. While SMD indices based on MORECS data can be usefully applied for regional assessment of risk to rail infrastructure, their application at site scale is not appropriate if the true variability relating to infiltration and recharge of groundwater into geotechnical asset is not captured. Whereas resistivity or moisture content images can be readily converted into images of SMD simply via knowledge of the fill resistivity at saturation. Indeed, the saturation index used in Fig. 7 provides a proxy to SMD (and it is highly likely that volumetric moisture content and saturation are more valuable indices for characterising the true distribution of internal conditions of earthworks assets).

DISCUSSION

Fundamentally, these case histories demonstrate that the construction method, deterioration history and distribution of composite materials within the embankment control engineering performance, especially, the spatial and temporal variation of groundwater and its influence upon key geotechnical properties controlling strength and stability. They relate to end members of a spectrum of engineered embankments. BIONICS, a modern clay embankment that was built up in layers, which are clearly recognised within the resistivity images. East Leake, an aged embankment with a heterogeneity that reflects the transition of a range of fill materials within an end tipped structure, again, identified on resistivity images. These case histories also demonstrate the application of time lapse resistivity images in understanding the link between weather events and subsurface processes and property changes affecting stability, which if applied to a 'virtual asset' provide the potential for predictive maintenance, for example within the context of resilient infrastructure in future climates.

The BIONICS example provides insight into how seasonal, cyclic wetting and drying in the near surface could drive the development of fissure networks deeper into the flank. For example, one can envisage similar processes driving the progressive moisture-driven cyclic strains and development of zones of low shear strength that could comprise long term instability due to shear failure. In addition to this holistic visualisation of the subsurface driving processes, recent innovations in time lapse, differential resistivity image processing now enable automated systems to track the movement of the individual sensors within the monitoring network [37] and [38]. We can now establish cause and effect between coupled subsurface and surface processes in rapid ground failure events. While we have not yet applied this technique to our engineered embankment sites, we have monitored up to 1.6 m of down slope movement on sensor groups at the top of a natural earth flow lobe with sixteen measurements over one year (inset in Fig. 11). This example relates to a landslide monitoring site near Malton, North Yorkshire where we have imaged the movement and break up of prograding earth flow lobes transporting reworked Whitby Mudstone over the underlying Staithes Sandstone (Fig. 11). The resistivity image clearly maps out the flow of the Whitby Mudstone (blues–greens) over the underlying Staithes Sandstone (yellows–reds).

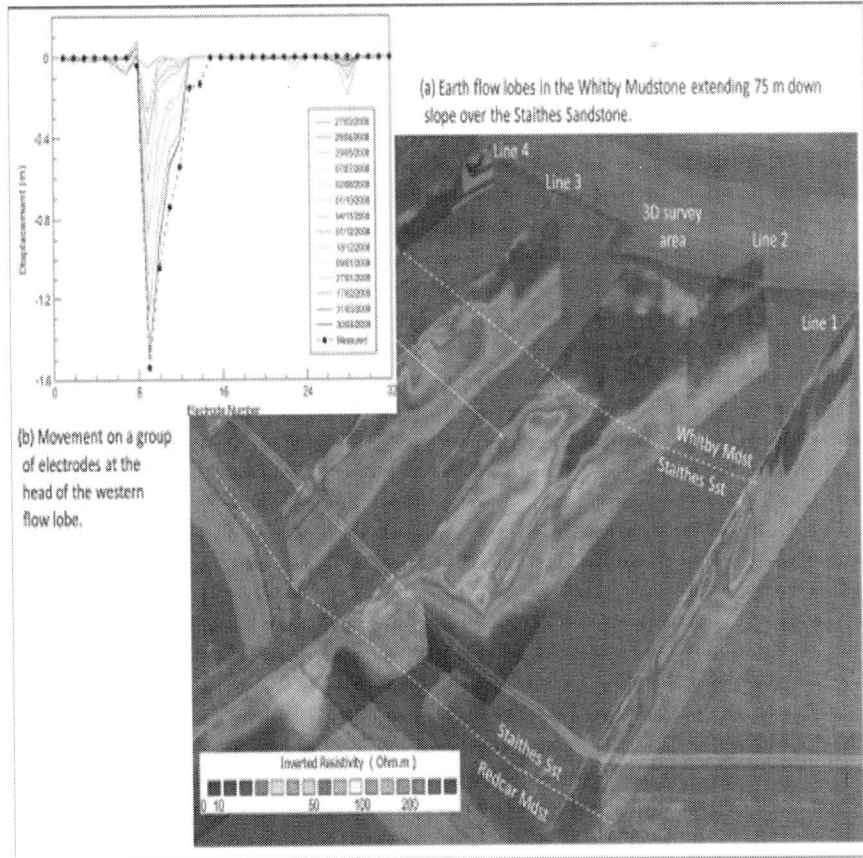

(a) Earth flow lobes in the Whitby Mudstone extending 75 m down slope over the Staithes Sandstone.

(b) Movement on a group of electrodes at the head of the western flow lobe.

Figure 11: Resistivity image and movement monitoring on earth flow landslide.

So, how could we apply these technologies to the management of earthworks assets? We can envisage programmable sequenced images resulting from remote field monitoring on permanent or semi-permanent installations telemetered and stored on a central database that contains multiple sites. These images would be processed to provide 2D, 3D and in time-lapse mode 4D images of resistivity, which are then transformed using robust relationships into geotechnical property ranges (moisture content, pore pressure, etc.). These property distributions would then be interrogated, for example to identify and classify internal infrastructure regimes. There can be multiple threshold levels set, for example, moisture content ranges could, for the basis of identifying regimes, include:

i.   *Dry*: below Shrinkage Limit, [Serviceability Limit State (SLS) – Monitor Subsidence].

ii.   *No Warning*: between Shrinkage and Plastic Limits.

iii. *Inspect*: between Plastic and Liquid Limits, [SLS – Monitor moisture levels; design drainage].

iv.   *Wet*: above Liquid Limit, [Ultimate Limit State – install drainage systems].

This information could form the basis for new proactive 'On-Demand' asset inspection scheduling. Automatic alarms could be programmed into the system that would issue a series of traffic light warnings that trigger certain actions. For example, the potential for shrinkage and crack formation in those zones triggering a 'Dry' warning could be monitored, either using independent systems or the resistivity based electrode movements. The subsurface image sequences could also map the rate and extent of drying and subsidence for use in considering the effect of vegetation and planning remedial actions (pollarding, etc.). As another example, zones triggering an 'Inspect' warning may indicate potential for plastic deformation (ballast pocket formation, mud pumping or even shear failure, etc.). Again, resistivity images would assist in monitoring moisture levels, movement pathways, sumps, springs, etc. and the subsequent design and scheduling of drainage measures. The detailed visualisation of subsurface ground water movement will enable design of lower cost measures that are better adapted to the specific causes of the problems – these are the subsurface processes. A consequence of responsive mode maintenance strategies is the use of surface based observations that currently define or diagnose the problem via the surface manifestations of the true cause, which is in fact, driven by the subsurface processes. By adopting subsurface imaging technology, we not only better define the cause of the problem (c.f. our use of 'CAUSE' in the East Leake case history hopefully captured the reader's attention), but more importantly, we also buy back a significant period of time in which to consider our early intervention. This is because the true cause of the problem begins as a progressive subsurface process, which could be very manageable if the subsurface symptoms are identified and enable robust prediction of future consequences, sufficient for appropriate preventative actions to be taken. All of this time advantage is lost by waiting for symptomatic surface manifestations, which are usually observations of direct surface movement, and hence maintenance is completely responsive, dealing only with the effect. Currently, the lack of understanding of the information potentially available from subsurface monitoring presents a key barrier inhibiting the development of preventative maintenance. Note also, responsive approaches may not address the true cause of the problem that if untreated, is left to continue to cause the problems. In this way we become trapped in a cycle of responsive maintenance, increasing whole life cycle costs, and quite possibly contributing to a reduction in the total lifespan of the asset.

Finally, by way of conclusion we provide a concept for a possible future linear route warning system. We envisage simple line arrays of electrodes, spaced between 2 m and 10 m extending along a 200 m–1 km long embankment. Two separate sets of monitoring measurements are made along the Up and Down line arrays. We have the capability to produce sequenced resistivity sections along these linear arrays that provide moisture content levels within depth intervals in the lower embankment, embankment core and possibly the upper embankment, again with independent sections for either flank of the asset. Under normal operating conditions, moisture level information is updated on a routine monitoring schedule, and these data are interrogated against a series of threshold classes such as the four above. The same routine monitoring schedule is maintained until certain weather event sequences trigger an 'Inspect' along a specific section (see Fig. 12). The warning also indicates the depth interval over which the high moisture levels occur, hence guiding impromptu inspection. This alarm also triggers a reconfiguration of the monitoring schedule, for example reducing the time interval between measurements for the purposes of studying potential temporal and spatial evolution of increased moisture levels. Having established the aerial and depth extent of the affected area, a scheme for drainage intervention is designed and scheduled for installation. Months (years?) later, the maintenance crews are mobilised for installation who don't observe any major surface expressions of distress but note seeping from pit walls at the intervals where the drainage measures are to be installed. By this time, the Inspect warning has been upgraded to 'Wet' – at ultimate limit state condition.

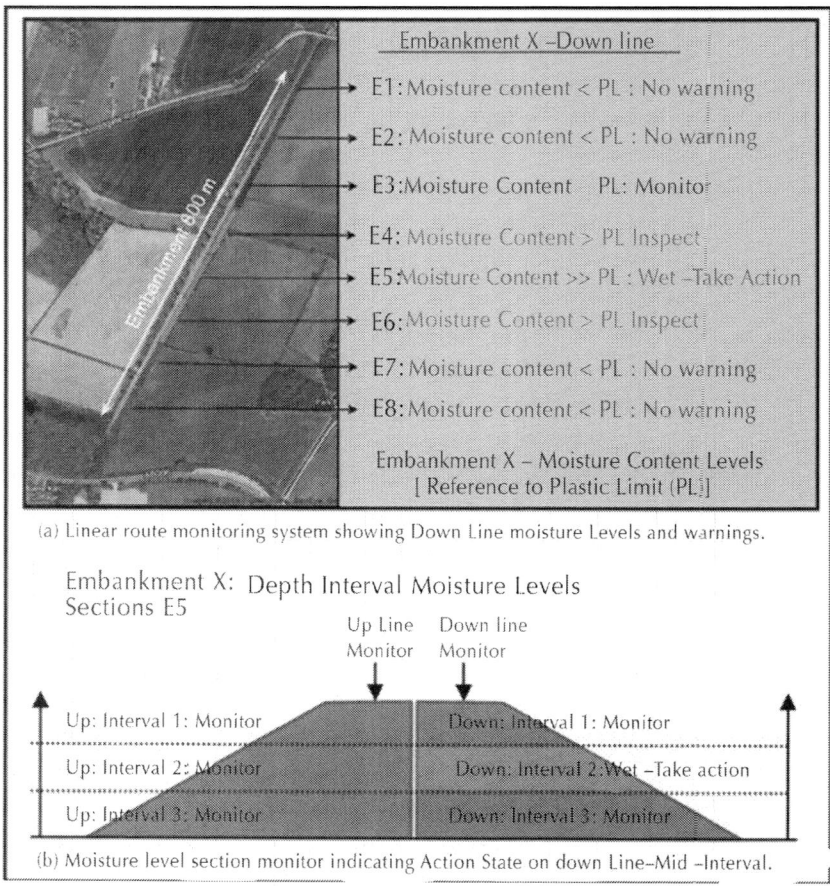

Embankment X –Down line

E1 : Moisture content < PL : No warning

E2 : Moisture content < PL : No warning

E3 : Moisture Content   PL : Monitor

E4 : Moisture Content > PL Inspect

E5 : Moisture Content >> PL : Wet –Take Action

E6 : Moisture Content > PL Inspect

E7 : Moisture content < PL : No warning

E8 : Moisture content < PL : No warning

Embankment X – Moisture Content Levels
[Reference to Plastic Limit (PL)]

(a) Linear route monitoring system showing Down Line moisture Levels and warnings.

Embankment X: Depth Interval Moisture Levels
Sections E5

Up Line   Down line
Monitor   Monitor

Up: Interval 1: Monitor   Down: Interval 1: Monitor

Up: Interval 2: Monitor   Down: Interval 2: Wet –Take action

Up: Interval 3: Monitor   Down: Interval 3: Monitor

(b) Moisture level section monitor indicating Action State on down Line–Mid –Interval.

Figure 12: Concept for a moisture monitoring system based upon real-time resistivity measurements over linear arrays along the Up and Down lines.

ACKNOWLEDGEMENTS

This paper is published with the permission of the Executive Director of the British Geological Survey (NERC). The authors gratefully acknowledge the Great Central Rail (Nottingham) Ltd. for allowing access to site and the publication of this manuscript.

REFERENCES

1. G.E. Archie, The electrical resistivity log as an aid in determining some reservoir characteristics, Petroleum Trans AIME, 146 (1942), pp. 54–62.

2. Bidder FW. The great central railway extension: Northern division. ICE, Vol. CXLII, Session 1899–1900, Part IV, Paper 3227, 1900. p. 1–22.

3. Bouch C, Jaroszweski D, Baker C, Dijkstra T, Dingwall R, Dixon N, Ryley T, Sivell P, Gunn D, Lawley, R. Future Resilient Transport Networks (FUTURENET): an overview of the FUTURENET project with particular reference to railway aspects. World Congress on Rail Research, 22–26 May 2011, Lille.

4. Bouch C and others incl., Gunn DA, Lawley R. Future resilient transport networks (FUTURENET): assessing transport network security in the face of climate change. In: 91st annual meeting of the transportation research board, January 22–26, 2012, Washington, DC.

5. Brough M, Stirling A, Ghataora G, Madelin K. Improving railway subgrade stiffness – assessment of traditional in situ ground improvement techniques. In: Proc. 3rd Int. Con. Railway Engineering, London; 2000.

6. P. Brunet, R. Clement, C. Bouvier, Monitoring soil water content and deficit using Electrical Resistivity Tomography (ERT) – a case study in the Cevennes area, France, J Hydrol, 380 (2010), pp. 146–153.

7. G. Cassiani, A. Godio, S. Stocco, A. Villa, R. Deiana, P. Frattini, et al., Monitoring the hydrologic behaviour of a mountain slope via time-lapse electrical resistivity tomography, Near Surf Geophys, 7 (2009), pp. 475–486.

8. Chambers JE, Gunn DA, Wilkinson PB, Ogilvy RD, Ghataora GS, Burrow MPN, Tilden Smith R. Non-invasive time-lapse imaging of moisture content changes in earth embankments using Electrical Resistivity Tomography (ERT). In: Ed. Ellis E, Yu HS, McDowell

G, Dawson A., Thom N, editors. Advances in Transportation Geotechnics. Proc. 1st int. conf. transportation geotechnics, Nottingham, August 2008. p. 475–480.

9.   Chambers JE, Gunn DA, Weller AL, Kuras O, Wilkinson PB, Meldrum PI, Ogilvy RD, Jenkins GO, Gibson AD, Ford JR, Price SJ. Geophysical anatomy of the Hollin Hill Landslide, North Yorkshire, UK. In: 14th European meeting of environmental and engineering geophysics, Krakow, Poland, 15–17 September, 2008.

10.   Chambers JE, Gunn DA, Meldrum PI, Wilkinson PB, Ogilvy RD, Haslam E, Holyoake S, Wragg J. Volumetric monitoring of earth embankment internal structure and moisture movement as a tool for condition monitoring. In: Proc. 11th int. conf. railway engineering, London; 2011.

11.   J.E. Chambers, P.B. Wilkinson, O. Kuras, J.R. Ford, D.A. Gunn, P.I. Meldrum, *et al.*, Three-dimensional geophysical anatomy of an active landslide in Lias Group mudrocks, Cleveland Basin, UK, Geomorphology, 125 (2011), pp. 472–484.

12.   J.E. Chambers, D.A. Gunn, P.B. Wilkinson, P.I. Meldrum, E. Haslam, S. Holyoake, *et al.*, 4D electrical resistivity tomography monitoring of soil moisture dynamics in an operational railway embankment, Near Surf Geophys, 12 (1) (2014), pp. 61–72.

13.   Fox FD. The great central railway extension: Southern division. ICE, Vol. CXLII, Session 1899–1900, Part IV, Paper 3209, 1900, p. 23–48.

14.   Glendinning S, Rouainia M, Hughes P, Davies O. Biological and engineering impacts of climate on slopes (BIONICS): the first 18 months. Int. Assoc. Eng. Geol., Nottingham, Paper 348; 2006.

15.   Glendinning S, Loveridge F, Starr–Keddle RE, Bransby MF, Hughes PN. Role of vegetation in sustainability of infrastructure slopes. In: Proceedings of the institution of civil engineers, engineering sustainability, 2009, 162, No. 2, p. 101–110.

16.   Gunn DA, Reeves H, Chambers JE, Pearson SG, Haslam E, Raines MR, Tragheim D, Ghataora G, Burrow M, Weston P, Thomas A, Lovell JM, Tilden Smith R, Nelder LM. Assessment of embankment condition using combined geophysical and geotechnical surveys. In: Proc. 9th int. conf. railway engineering, London; 2007.

17. Gunn DA, Reeves H, Chambers JE, Ghataora G, Burrow M, Weston P, Lovell JM, Tilden Smith R, Nelder LM, Ward D. New geophysical and geotechnical approaches to characterise under utilised earthworks. In: Proc. 1st int. conf. transportation geotechnics, Nottingham; 2008.

18. Gunn DA, Haslam E, Kirkham M, Chambers JE, Lacinska A, Milodowski A, Reeves H, Ghataora G, Burrow M, Weston P, Thomas A, Dixon N, Sellers R, Dijkstra T. Moisture measurements in an end-tipped embankment: Application for studying long term stability and ageing. In: Proc. 10th int. conf. railway engineering, London; 2009.

19. Gunn DA, Raines MG, Chambers JE, Haslam E, Meldrum PI, Holyoake S, Kirkham M, Williams G, Ghataora GS, Burrow MPN. Embankment stiffness characterisation using MASW and continuous surface wave methods. In: Proc. 11th int. conf. railway engineering, London; 2011.

20. Hughes P, Glendinning S, Mendes J. Construction Testing and Instrumentation of an infrastructure testing embankment. In: Proc. expert symposium on climate change: modelling, impacts & adaptations, Singapore; 2007. p. 159–66.

21. Hughes PN, Glendinning S, Davies O, Mendes J. Construction and monitoring of a test embankment for evaluation of the impacts of climate change on UK transport infrastructure. In: Proc. 1st int. conf. transportation geotechnics, Nottingham; 2008.

22. Hughes PN, Glendinning S, Mendes J, Parkin G, Toll DG, Gallipoli D, Miller P. Full-scale testing to assess climate effects on embankments. Special issue of engineering sustainability, Institution of Civil Engineers, 162, No. ES2, 2009, p. 67–79.

23. IPCC. Intergovernmental panel on climate change: synthesis report. In: Pachauri RK, Reisinger A, editors. Contribution of working Groups I, II and III to the fourth assessment report. Cambridge Univ. Press: Cambridge, UK; 2007. 104p.

24. IPCC. Climate Change 2007: Impacts, Adaptation and Vulnerability. In: Parry ML, Canziani OF, Palutikof JP, van der Linden PJ, Hanson CE, editors. Contribution of working Group II to the Fourth assessment report of the intergovernmental panel on climate change. Cambridge University Press: Cambridge, UK; 2007. 976 p.

25. Li D, Selig ET. Evaluation of railway subgrade problems. transportation research Record No. 1489, Rail, Transportation Research Board, Washington, DC; 1995. p. 17–25.

26. Loke MH. RES2DINV Rapid 2-D resistivity and IP inversion using the least-squares method. Manual, Geotomo Software, Penang, Malaysia; 2006.

27. Niesner E. Subsurface resistivity changes and triggering influences detected by continuous geoelectrical monitoring. The Leading Edge, August 2010. p. 952–55.

28. O'Brien AS. Rehabilitation of urban railway embankments – investigation, analysis and stabilisation. In: Cuellar V, Dapene E et al., editors. Proceedings of XIV European conference on soil mechanics and geotechnical engineering, Madrid, September 2007.

29. Ogilvy RD, Meldrum PI, Kuras OA, Wilkinson PB, Chambers JE. Advances in geoelectrical imaging technologies for the measurement and monitoring of complex earth systems and processes. In: 33rd int. geol. congress, Oslo; 2008.

30. R.D. Ogilvy, P.I. Meldrum, O. Kuras, P.B. Wilkinson, J.E. Chambers, M. Sen, et al., Automated monitoring of coastal aquifers with electrical resistivity tomography, Near Surf Geophys, 7 (2009), pp. 367–375.

31. Okada K, Ghataora GS. Assessment of the stiffness of railway subgrade. In: Proc. 3rd int. con. railway engineering, London; 2000.

32. Reeves GM, Sims I, Cripps JC. Clay materials used in construction. Geological Society Engineering Geology Special Publication No. 21; 2004. 517p.

33. Selig ET, Waters GM. Track geotechnology and substructure management, Thomas Telford; 1994. 446p.

34. V. Shevnin, A. Mousatov, A. Ryjov, O. Delgado-Rodriquez, Estimation of clay content in soil based on resistivity modelling and laboratory measurements, Geophys Prospect, 55 (2007), pp. 265–275.

35. P. Sjodahl, T. Dahlin, S. Johansson, Using the resistivity method for leakage detection in a blind test at the Rossvatn embankment

dam test facility in Norway, Bull Eng Geol Environ, 69 (2010), pp. 643–658.

36. M.H. Waxman, L.J.M. Smits, Electrical conductivities in oil-bearing shaly sands, Soc Petrol Eng J, 8 (1968), pp. 107–122.

37. P.B. Wilkinson, J.E. Chambers, P.I. Meldrum, D.A. Gunn, R.D. Ogilvy, O. Kuras, Predicting the movements of permanently installed electrodes on an active landslide using time-lapse geoelectrical resistivity data only, Geophys J Int, 183 (2010), pp. 543–556.

38. P.B. Wilkinson, J.E. Chambers, O. Kuras, P.I. Meldrum, D.A. Gunn, Long-term time-lapse geoelectrical monitoring, The First Break, 29 (2011), pp. 77–84.

11

Prevention and Treatment Technologies of Railway Tunnel Water Inrush and Mud Gushing in China

Yong Zhao[a], Pengfei Li[b], and Siming Tian[a]

[a]Project Design and Approval Center of Ministry of Railways, Beijing 100844, China
[b]College of Architecture and Civil Engineering, Beijing University of Technology, Beijing 100022, China

ABSTRACT

Water inrush and mud gushing are one of the biggest hazards in tunnel construction. Unfavorable geological sections can be observed in almost all railway tunnels under construction or to be constructed, and vary in extent. Furthermore, due to the different heights of mountains and the lengths of tunnels, the locations of the unfavorable geological sections cannot be fully determined before construction, which increases the risk of water inrush and mud gushing. Based on numerous cases of water inrush and mud gushing in railway tunnels, the paper tries to classify water inrush and mud gushing in railway tunnels in view of the conditions of the surrounding rocks and meteorological factors associated with tunnel excavation. In addition, the causes of water inrush and mud gushing in combination of macro- and micro-mechanisms are summarized, and site-specific treatment method is put forward. The treatment methods include choosing a method of advance geological forecast according to risk degrees of different sections in the tunnel, determining the items of predictions, and choosing the appropriate methods, i.e. draining-oriented method, blocking-oriented method or draining-and-blocking method. The treatment technologies of railway water inrush and mud gushing are also summarized, including energy relief and pressure relief technology, advance grouting technology, and advance jet grouting technology associated with their key technical features and applicable conditions. The results in terms of treatment methods can provide reference to the prevention and treatment of tunnel water inrush and mud gushing.

INTRODUCTION

Up to now, nearly ten thousand kilometers of railway tunnels have been built under a variety of complex geological conditions in China. In tunnel construction, water inrush and mud gushing are the challenging issues. They are characterized by the largest geological hazards and the potential risks (see Fig. 1). In this paper, water inrush and mud gushing in underground projects basically refer to the dynamic destruction of structure in unfavorable geological sections. The hydrodynamic system and the dynamic equilibrium of surrounding rocks undergo drastic changes due to the excavation of underground projects, which causes instant release of the energy stored in underground water carrying mud and sand that gush to the tunnel face at a high speed (Lin and Song, 2012).

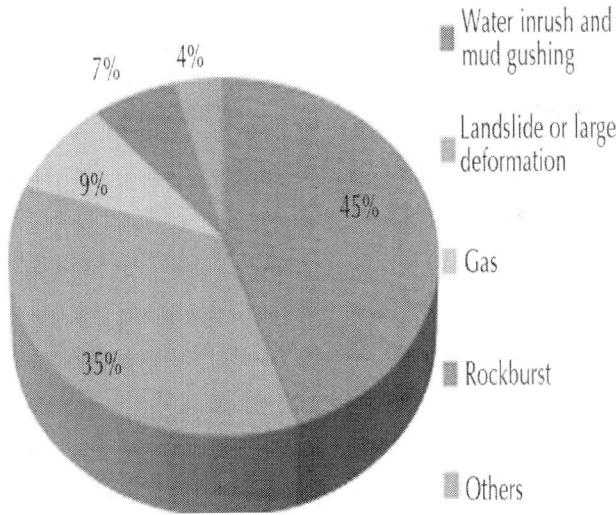

Figure 1: Types of geological disasters in tunnels excavated by drill-and-blast method and their percentages.

In recent years, more than 100 cases of water inrush and mud gushing have been observed in China, causing serious losses of human life and economics and deterioration of construction conditions. Accordingly, the quality and the long-term safety of underground construction are involved. For drainage in large quantity over a long period of time is costly, the ecological environments can seriously be undermined, leading to the depletion of water resources, death of vegetation and destruction of farmlands.

Water inrush seriously threatens construction when the tunnel goes through water-rich area, which frequently occurs in unfavorable geological sections. For examples, serious water inrush and mud gushing occurred in the Yuanliang mountain tunnel of Chongqing-Huaihua Railway. The peak duration of the water inrush lasted for roughly 0.5 h, with water discharge of 1.1×10^5 m^3; the mud in the tunnel was 1.2–4.5 m deep, and the value of mud was approximately 1.5×10^4 m^3. Large-scale events of water inrush and mud gushing of this kind occurred ten times in Wulong mountain tunnel, accounting for economic losses over RMB 20 million. In the Malujing tunnel of Yichang-Wanzhou Railway, water inrush and mud gushing occurred in the "DK255 + 978" karst tunnels on 21 January 2006, with the value of water inrush of approximately 7.2×10^5 m^3 and the mud of about 7.0×10^4 m^3. The water inrush and mud gushing flooded the 3152 m parallel heading and 2508 m main tunnel in vicinity of exit, and caused the damages of a large number of equipment and machinery with economic losses over RMB 10 million. After that, the treatments of this accident took for 3 years. In the Yesanguan tunnel of Yichang-Wanzhou Railway, water inrush and mud gushing were observed in "DK124 + 602" karst tunnels on 5 August 2007, and the peak flow rate reached 1.5×10^5 m^3/h. The volume of mud and broken stones gushing was around 5.0×10^4 m^3, causing the equipment and machinery in the tunnel to be seriously deformed and/or damaged, flooded about 500 m away from their positions. The economic losses were huge and subsequent treatments took 2 years. In the Baiyun tunnel of Nanning-Guangzhou Railway, water inrush and mud gushing occurred at the stake DK334 + 733 on 16 January 2010, lasting for about 30 s. The length of mud gushing area is 167 m, and the value of mud and sand gushing is over 2500 m^3, also leading to serious damages to construction equipment and machinery. The Guanjiao tunnel of Xining-Lhasa Railway, located at the northeast edge of Qinghai-Tibet Plateau with strong geological activities, goes through 19 fault zones with the total length of 2.8 km. The peak water inrush rate reaches 3.2×10^5 m^3/d with water loss of 1.5×10^5 m^3/d, inducing the cost of electricity output RMB 5 million monthly.

In unfavorable geological sections, it is critically important to figure out the mechanism of water inrush for safe tunnel construction, and effective water inrush risk assessment method and appropriate measures should be adopted. At present, researches on the mechanism of water inrush in tunnel construction in mountainous regions, such as the analysis of water inrush mechanism in unfavorable geological sections (especially in fault zones and karst stratum), determination of critical condition of water inrush, the quantity calculation of inrush water, are fruitful, providing solid supports to tunnel construction (Guan, 2003, Cui, 2005, Jiang, 2006, Zhang, 2010 and Guan and Zhao, 2011). However, most researches on water inrush in tunnels are experienced from coal mining industry (Gao et al., 1999, Li, 2010 and Guo, 2011). For extra-long tunnels in mountains, the locations of the unfavorable geological sections cannot be fully understood before construction. We still do not have very effective method for advance geological forecast. We need to have better understanding of the disaster-causing mechanism and dynamic evolution of the disasters, and correct theories for the early warning and prevention of water inrush and mud gushing.

Based on the characteristics of long tunnels, this paper summarizes different types of water inrush and mud gushing accidents and tries to understand the influential factors and propose preventive measures and treatment technologies which have been successfully applied to many railway tunnels in China.

THE MECHANISMS AND INFLUENTIAL FACTORS OF WATER INRUSH AND MUD GUSHING IN TUNNELS

Types of Water Inrush and Mud Gushing

Water inrush and mud gushing in tunnels can be classified into different types according to various criteria. Based on the results (Gao et al., 1999, Zhao et al., 2009, Li, 2010 and Guo, 2011), the classification of water inrush and mud gushing is shown in Table 1 and Fig. 2.

Table 1: Types of water inrush and mud gushing

Classification criteria	Type
The location of water inrush and mud gushing	(1) Water inrush and mud gushing at tunnel face
	(2) Water inrush and mud gushing at tunnel vault
	(3) Water inrush and mud gushing at tunnel floor
	(4) Water inrush and mud gushing at tunnel sidewalls
The occurring process of water inrush and mud gushing	(1) Sudden burst of water inrush and mud gushing
	(2) Delayed water inrush and mud gushing
	(3) Sluggish water inrush and mud gushing

The quantity of water inrush and mud gushing	(1) Super large scale water inrush and mud gushing
	(2) Large-scale water inrush and mud gushing
	(3) Medium scale water inrush and mud gushing
	(4) Small-scale water inrush and mud gushing
	(5) Mini-scale water inrush and mud gushing
The influential factors of water inrush and mud gushing	(1) The sudden breaking of karst cavity
	(2) The breaking of fault fracture zone of confined water
	(3) Fracture of channel and bedrock in areas of high water pressure and rich water
	(4) Mud gushing in regional water-rich cystic weathering cavities

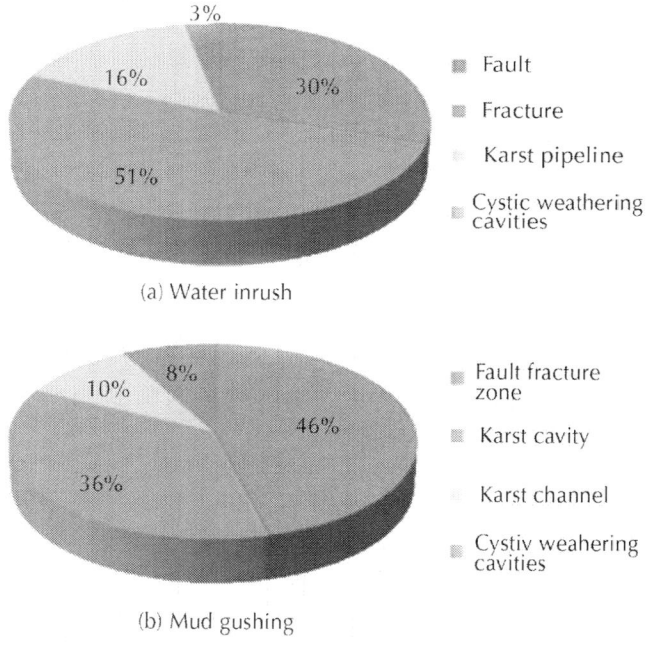

(a) Water inrush

- Fault
- Fracture
- Karst pipeline
- Cystic weathering cavities

(b) Mud gushing

- Fault fracture zone
- Karst cavity
- Karst channel
- Cystiv weahering cavities

Figure 2: Types of water inrush and mud gushing in tunnels and their percentages.

Influential Factors of Water Inrush and Mud Gushing

Through site investigation and available studies of water inrush and mud gushing accidents, it is observed that the tunnels with water inrush and mud gushing are basically characterized by high water pressure, large quantity of water storage, loose filling materials, rich water supply, and known geological conditions. The main factors, related to water inrush and mud gushing, are the sources of water supply, water pressure, confining bed, geological structure, tunnel excavation and meteorological impacts. Source of water supply is the dominant factor; water pressure is the main factor, a driving force determining whether water inrush occurs and the quantity thereafter; the unbroken rock stratum above the vault can be regarded as the confining bed to ensure safe tunnel excavation; geological structure determines the passage of the water inrush in tunnels (most water inrush cases especially that of large scale are related to geological structure); tunnel excavation is the inducing factor of water inrush; temperature, raining and other meteorological factors contribute to water inrush and mud gushing. The first four factors are associated with the conditions of surrounding rocks.

Conditions of Surrounding Rocks

The environmental conditions of tunnel excavation, such as the rock types and lithology, conditions of underground water and other geological factors, are the basic conditions for water inrush and mud gushing. Tunnel water inrush and mud gushing usually occur at the structural plane of the rocks, fault zone, intensively fissured zone, weathered trough, karst cavity and other unfavorable geological areas. In terms of rock lithology, water inrush and mud gushing are usually observed at dolomite, limestone and other dissolvable rock strata. The vertical zonation of water and mud is closely related to water inrush. The possibility of tunnel water inrush emergence from low to upper is karst water aeration zone, deep slow flow zone, seasonal change zone, shallow water-rich zone, and water-rich zone with pressure.

Impact of Raining

Raining, temperature changes and other meteorological phenomena can also affect water inrush and mud gushing, i.e. magnitude varying in different seasons. For example, in summer with strong rainfall, the construction-induced disturbance in combination of high water pressure and temperature, water scouring effect and other relatively actives (i.e. physico-chemical reactions) can easily induce large-scale water inrush and mud gushing.

Construction-Induced Disturbance

Tunnel excavation can inevitably disturb the original strata. Excavation method and excavation extent are two important factors. The former is reflected in whether drill-and-blast method or partial excavation method is adopted and in the design of support parameters; the latter is involved in cross-sectional area and excavation span.

Mechanism of Water Inrush and Mud Gushing

The macro-mechanism of tunnel water inrush and mud gushing is basically concentrated on the analysis of different types of tunnel water inrush and mud gushing, and the micro-mechanism on the analysis of the minor physico-mechanical effect of underground water exerted on the strata. Thus, analysis of macro-mechanism can explain different types of water inrush and mud gushing, and that of micro-mechanism can provide theoretical basis (Lin and Song, 2012) for an in-depth study on the causes of water inrush and mud gushing and the mechanism.

In karst strata, engineering geology, hydro-geological conditions, in situ stress level, and physico-mechanical characteristics of rocks as well as different tunnel locations, lead to different mechanisms of tunnel water inrush and mud gushing. Thus, the mechanism of emergence of tunnel water inrush and mud gushing is different due to complex geological conditions. In terms of macro-mechanical mechanism, tunnel water inrush and mud gushing can be classified into 4 types according to the storage environments of water and mud: karst cave (hole) water inrush, high-pressure geological interface water inrush, and underground karst channel (or underground river) water inrush, and fault (fractured) zone water inrush. According to analysis of formation of mud passage, the mechanism of tunnel water inrush can be roughly classified into 4 types: the fracture of watertight layer through stretching, the shear-yielding of geological interface, the hydraulic impact of discontinuous geological interface and the instability and slipping of the controlling rocks (Lin and Song, 2012).

The essence of micro-mechanism of tunnel water inrush is the continuous physico-chemical impact of water and mud on geological interface and the impact of the accumulated micro-damages, including the softening and dissolving effect of water and mud on the strata of the geological interface, the effect on cavity expansion, the water wedge effect, and the erosion-expansion effect on the water inrush passage. Therefore, tunnel water inrush is a process evolution in which the continuous physico-chemical impact of water and mud on geological interface leads to its progressive damage. The erosion-expansion effect of groundwater is shown in Fig. 3.

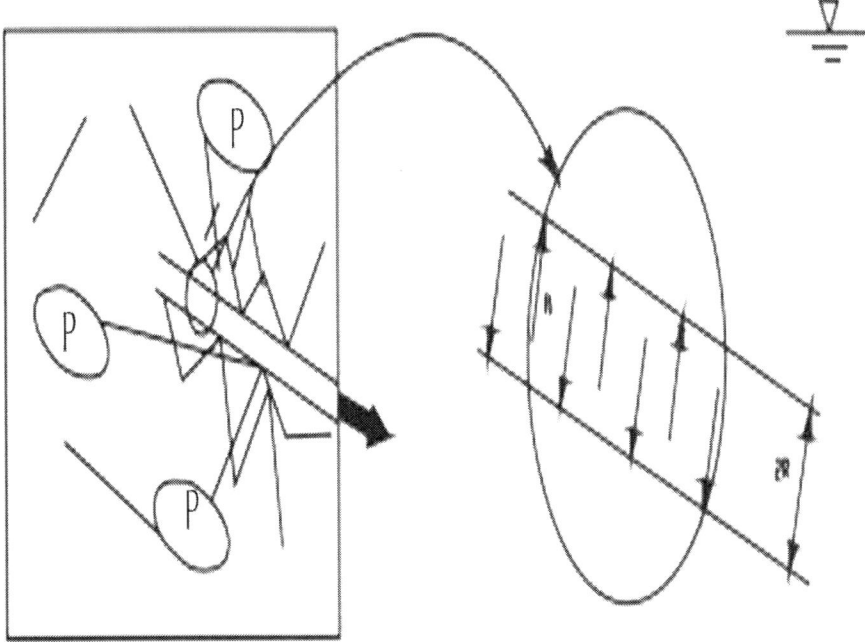

Figure 3: The erosion-expansion effect of groundwater.

PREVENTION MEASURES AND TREATMENT TECHNOLOGIES OF TUNNEL WATER INRUSH AND MUD GUSHING

Prevention Measures

Comprehensive advance geological forecast is the principal measure for the prevention of railway tunnel water inrush and mud gushing. Currently, it has been widely used in railway tunnel construction in China. The measures include geophysical prospecting and prospecting by drilling. Advance geological forecast shall be determined according to the risks of different sections in the tunnel, and the prediction items shall be considered subsequently. The classification of advance geological forecast and the items are shown in Table 2 (Zhao et al., 2009). According to the researches of Shandong University (group of Prof. Li Shucai), advance geological forecast technology (Li, 2009, Zhang et al., 2010 and Li et al., 2011) for progressive water-bearing structures is shown in Fig. 4.

Table 2: The classification of advance geological forecast and items to be forecasted

Classification	Geological features	Items to be predicted
A+	(1) High proneness of geological disaster areas, such as large underground river system, water-rich weak faults with good hydraulic conductivity. (2) Existence of major geophysical anomaly, most probable occurrence of water inrush and mud gushing at the rate of over 10,000 m^3/h, and most likely to induce major environmental and geological disasters.	(1) Geological sketch. (2) Long distance forecast: TSP203 (100 m). (3) Medium distance forecast: advance level hole drilling (30–60 m) 1–3 holes; if necessary, HSP or negative apparent velocity method with 50 m. (4) Short distance forecast: geological radar (15–30 m), continuous infrared water prospecting, advance level hole drilling (30 m) 6 holes; advance blast hole (5 m) 5 holes, and if necessary, use CT or borehole camera.
A	(1) Probable existence of geologically disastrous areas, such as large underground river system, weak fault rich in water and with good hydraulic conductivity. (2) Existence of major geophysical anomaly, most probable occurrence of water inrush and mud gushing at the rate of over 1000 m^3/h, and possibility to induce major environmental and geological disasters.	(1) Geological sketch. (2) Long distance forecast: TSP203 (100 m). (3) Short distance forecast: advance level hole drilling (30 m) 1–3 holes; advance blast hole (5 m) 3–5 holes; geological radar (15–30 m), if necessary, continuous infrared water prospecting.

| B | Medium water inrush and mud gushing at the rate of 100–1000 m³/h, existence of comparatively large geophysical anomaly, fault zone, etc. | (1) Geological sketch. (2) Long distance forecast: TSP203 (150 m). (3) Short distance forecast: advance level hole drilling (30 m) 1–3 holes; advance blast hole (5 m) 3 holes; geological radar (15–30 m) at geophysical anomaly, infrared water prospecting for 20 m. |
| C | Carbonate and clastic rocks with relatively good hydrogeological conditions, low probability of water inrush of mud gushing, or its rate less than 100 m³/h. | (1) Geological sketch. (2) TSP203 (150 m) for major geological interface or geophysical anomaly, advance blast hole (5 m) 3 holes. (3) If necessary, geological radar (30 m). |

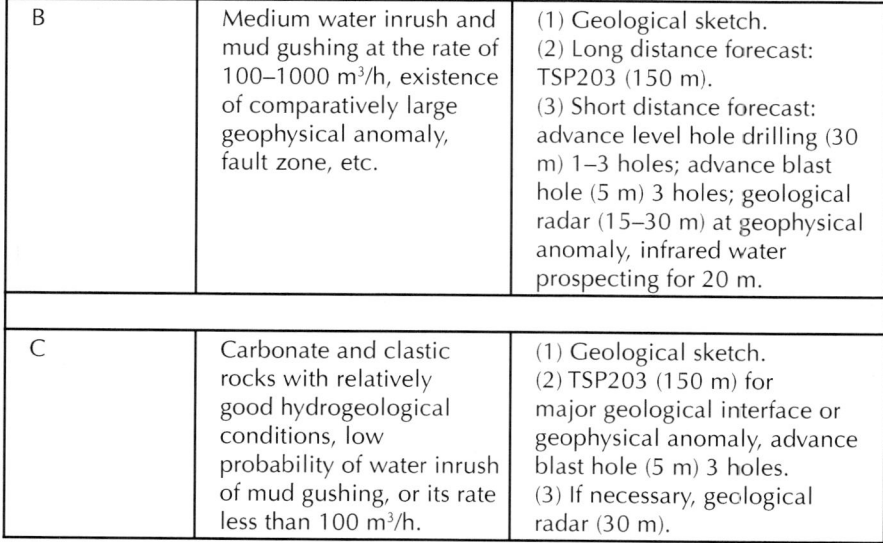

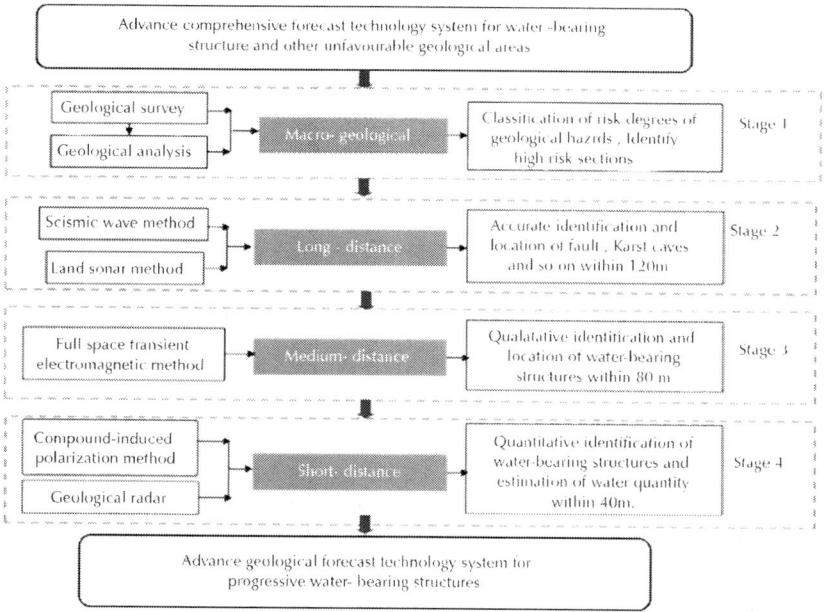

Figure 4: Advance geological forecast technology system for progressive water-bearing structures.

Treatment Principle and Technology

Treatment Principle

According to the site-specific groundwater environments, the treatment principles are draining-oriented, blocking-oriented or draining-and-blocking.

1. *Draining-Oriented*: such as energy relief and pressure reduction method. It is mainly applied to large-scale high-pressure water-rich karst, high-pressure and water-rich karst channel, and regional water-rich cystic weathering cavities.

2. *Blocking-Oriented:* such as advance curtain grouting method. It is mainly applied to the cases that groundwater environment requirement is strict, the faults where water pressure is less than 0.5 MPa, and the water leakage from bedrock fissure.

3. *Draining-and-Blocking:* such as pressure relief through water diversion tunnels, and grouting. It is mainly applied to water-rich faults where water pressure is over 0.5 MPa and water leakage from bedrock fissure, or karst deposits after energy relief and pressure reduction.

Treatment Technology

(1)Energy Relief and Pressure Reduction Method

Energy relief and pressure reduction method (Zhang, 2010) means in karst cavity with high pressure and rich water during tunnel construction, accurate blasting or advance drilling is applied to drain water, release mud, and reduce water pressure in the cavity for the purpose of energy relief. During excavation, structural support and treatment after energy relief and pressure reduction can take a certain period of time. Energy relief and pressure reduction method shall be considered in dry season; if it is used in rainy season, the constant supply of rainwater can pose risk to tunnel construction. The work procedure and detailed items of energy relief and pressure reduction method are shown in Fig. 5.

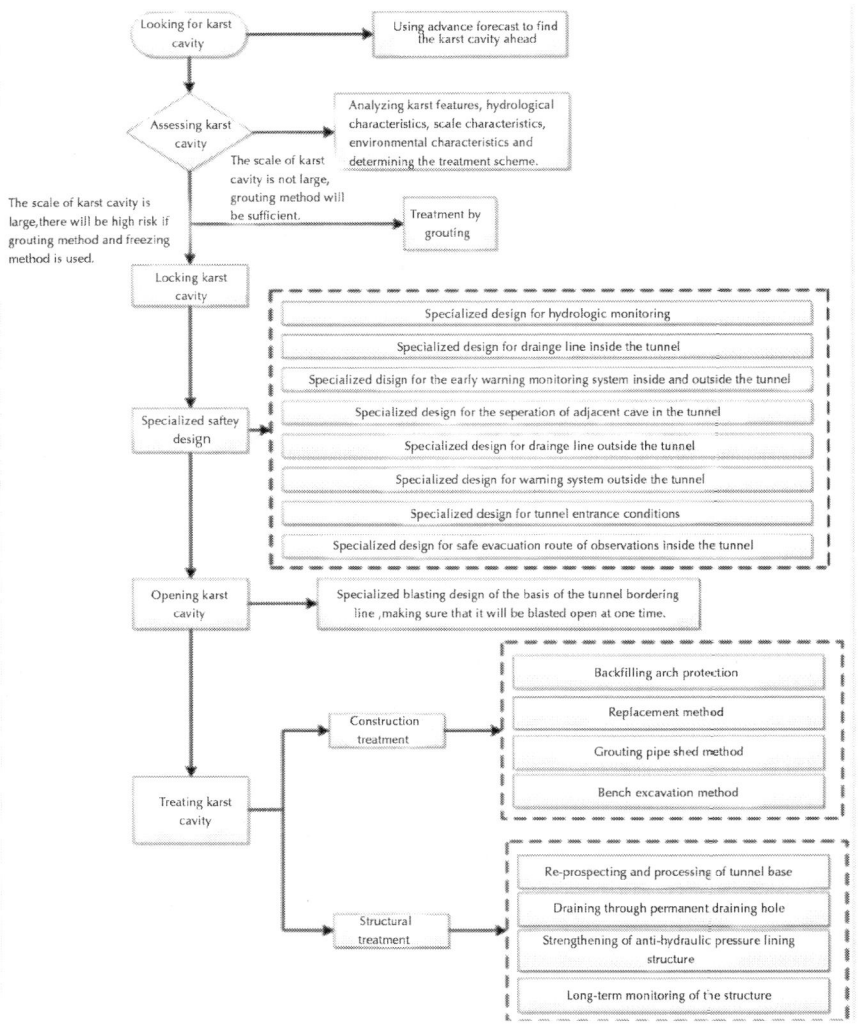

Figure 5: Work procedure and description of energy relief and pressure reduction method.

(2)Advance Grouting Technology

Advance grouting means filling proper grouting materials into karst cavity, fault (fractured) zone, bedrock fissure and other construction bodies through reasonable grouting process by equipment and machinery, so as to achieve filling, reinforcement, water-blocking to ensure the safe tunnel excavation and its long-term operation (Mo and Zhou, 2008, Dai, 2009 and Zhuang and Mu, 2009).

Commonly used advance grouting technology includes full-face curtain grouting technology and informationized grouting technology. Full-face curtain grouting technology is based on the assumption that the outer strata of the tunnel are uniform, thus the loose area can be reinforced by full-face grouting. In practical engineering, grouting parameters can be determined according to water pressure as shown in Fig. 6. For informationized grouting technology, according to different damages of the strata, sectional division, water blockage from outside and reinforcement inside, local strengthening is used as shown in Fig. 7. Compared with the full-face curtain grouting technology, informationized grouting technology involves few drilling holes and thus is less time-consuming.

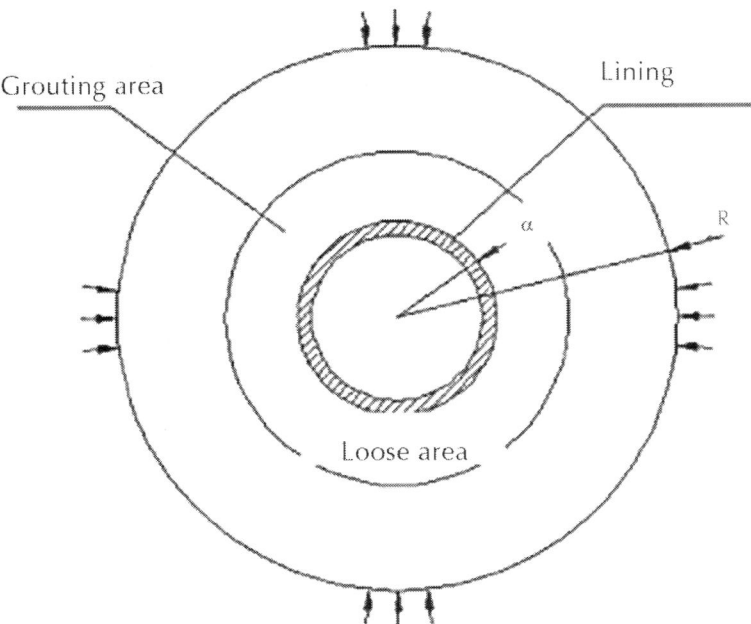

Figure 6: Schematic diagram of full-face curtain grouting.

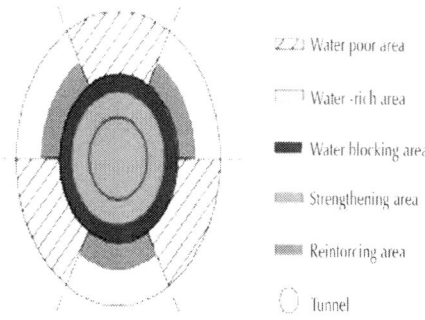

Figure 7: Schematic diagram of informationized grouting.

(3) Advance Jet Grouting Technology

Advance jet grouting technology (Yang and Zhang, 2008 and Zhao, 2012) is basically employed for the purpose of deformation control of surrounding rocks. Through measuring core rock and soil samples, the physico-mechanical properties of surrounding rocks can be determined; accordingly, the states of the surrounding rocks, i.e. stable, temporary stable, unstable after excavation, can be predicted. Then, the informationized design and construction method can be considered to control the geological deformation and to ensure safe passing of tunnel's full-section in various strata, in especially complex and unfavorable geological regions. Presently, the technology is introduced to the treatment of railway tunnel water inrush and mud gushing (see Fig. 8).

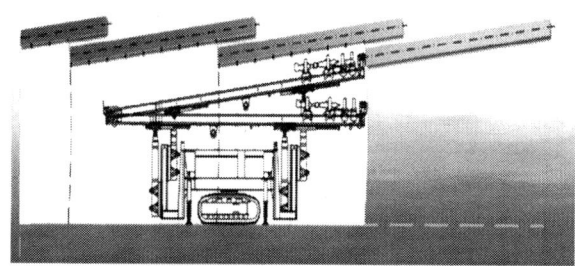

Figure 8: Diagram of advance level jet grouting in tunnels.

APPLICATIONS OF TREATMENT TECHNOLOGIES

Application of Energy Relief and Pressure Reduction Method to Yesanguan Tunnel

After the occurrence of water inrush and mud gushing in Yesanguan tunnel of Yichang-Wanzhou Railway, it is suggested that the energy relief and pressure reduction method and grouting method be adopted. A drainage hole of 5250 m long is considered on the left side of the tunnel, and a branch drainage hole at a high position, 7.5 m above the main tunnel vault, is dug to release energy and reduce water pressure (see Fig. 9). Accordingly, the mixture of water and mud in the cave was released to mitigate the construction risks. After that, advance pipe-shed and other measures are adopted, and finally the tunnel construction is successfully completed. Three years' monitoring results show that the tunnel structure is safe and reliable.

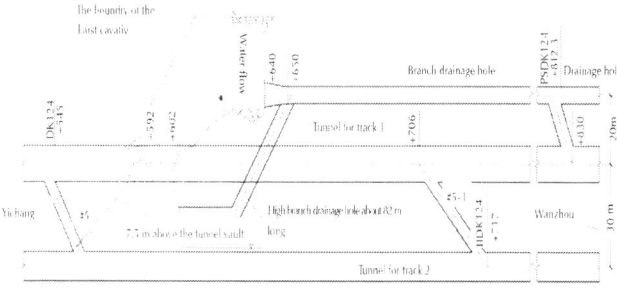

Figure 9: Plane sketch of the treatment of "+602" karst cavity.

Application of Curtain Grouting Method to Qiyue Mountain Tunnel

Qiyue Mountain tunnel is located at the junction of soluble rock and insoluble rock, passing through the fault F11 characterized by high water pressure. The fault F11 is large-scale stretching about 250 m along the tunnel direction, composed of limestone, tectonic breccia, fault gouges. The rock is broken and poor in agglutination. Water inrush rate of a single hole disclosed by advance geological drilling reaches 790 m³/h. The water pressure is observed up to 2.5 MPa. Before tunnel construction, the fault F11 is carefully investigated by comprehensive advance geological forecast methods, such as geological sketches, advance geophysical prospecting, 60 m long advance hole drilling, 30 m long advance hole drilling, and blast hole. As a result, large-scale water inrush and mud gushing are successfully prevented.

The principle and method of "pressure relief through draining + reinforcement by grouting" are adopted according to the obtained information by the advance geological forecast method and field grouting and draining tests. Drainage holes are considered in the parallel heading and branch drainage areas to relieve water pressure of the surrounding areas in the tunnel, thus water pressure on the surrounding rocks of the tunnel is reduced, and consequently the difficulties in tunnel grouting and excavation are reduced (see Fig. 10). With those measures, the scopes of the parallel heading and the main tunnel by advance grouting are determined at 5 m and 8 m outside the excavation contour line, respectively (see Fig. 11). Additional site-specific grouting holes shall also be considered in water-rich regions. In this project, the advance pipe shed and tunnel face fiberglass bolt and other advance-supporting measures are taken for consideration of safe and quick tunnel excavation. After 3 years of monitoring, it shows that the tunnel structure remains safe and reliable.

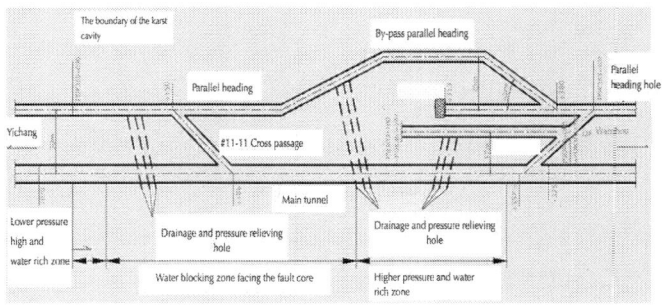

Figure 10: Plane sketch of the geological features and treatment method of fault F11 in Qiyue Mountain tunnel.

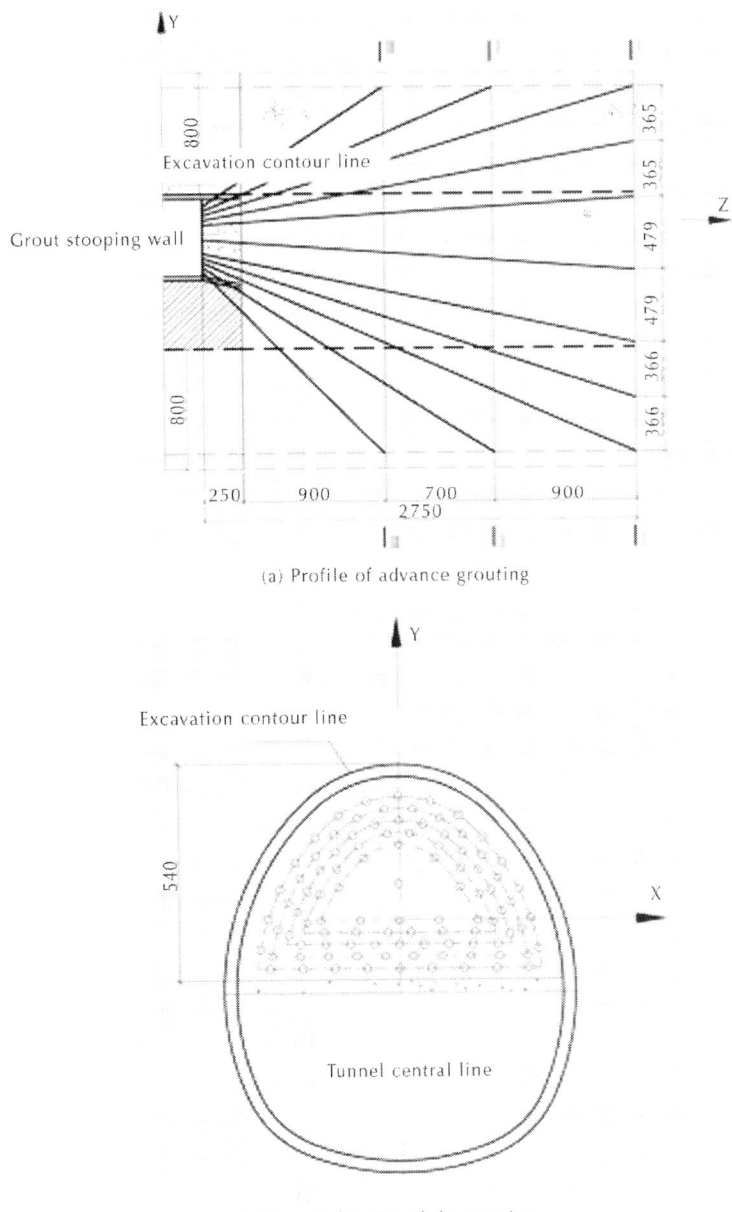

(a) Profile of advance grouting

(b) Layout drawing of the opening

Figure 11: Design of curtain grouting in Qiyue Mountain tunnel (unit: cm).

Application of Informationized Grouting Method to Guanjiao Tunnel

Guanjiao tunnel in the Qinghai-Tibet Railway is located between Tianpeng station (the existing line) and Chahannuo station. There are two parallel tracks in the tunnel with length of 32.6 km (currently the longest railway tunnel in Asia). The tunnel passes through many water-rich fault zones and serious water inrush disasters occurred during construction. In order to reduce the quantity of water inrush and the costs of water drainage, the informationized grouting technology is adopted. The number of necessary grouting holes is determined considering the results of advance geological forecast, damage extent of surrounding rocks at the tunnel face, water inrush quantity and forms of water inrush. The principle of water inrush control is applied from outside to inside, step by step, through drilling at suitable intervals. Addition of holes and local strengthening are also considered to achieve the targeted number of grouting holes and the effective grouting. The informationized grouting technology adopted in the construction of Guanjiao tunnel is shown in Table 3, and good results are achieved (see Fig. 12).

Table 3: Types of informationized grouting methods employed in Guanjiao tunnel

Types of grouting	Grade of surrounding rocks	Criteria of initiation grouting	Criteria of grouting termination
Radial grouting after excavation	Relatively good surrounding rocks (such as rock classes II and III)	Rate of water inrush in a single hole of excavated area is 5–40 m³/h, existence of many water inrush points	The quantity of water inrush in the grouted section does not exceed 10 m³/(m d)
Local water inrush by grouting	Local water inrush around the tunnel, mainly surrounding rocks classes III and IV	Rate of water inrush around the tunnel in a single hole is over 40 m³/h	The rate of water inrush in a single hole in the concentration point of water inrush is less than 10 m³/h

Advance grouting	Local water inrush in tunnel bed, mainly surrounding rocks classes III and IV	Rate of water inrush in a single hole of the tunnel face is over 40 m³/h	The rate of water inrush in a single hole in the concentration point of water inrush is less than10 m³/h, and the total rate of water inrush at the tunnel face is less than 100 m³/h
Periphery curtain grouting	Weak and broken surrounding rocks classes IV and V		
Full-face curtain grouting	Fault, fractured surrounding rocks classes V and VI	The rate of water inrush in a single hole of the tunnel face is over 40 m³/h	The rate of water inrush in a single hole in the concentration point of water inrush is less than 5 m³/h, and the total rate of water inrush at the tunnel face is less than 100 m³/h

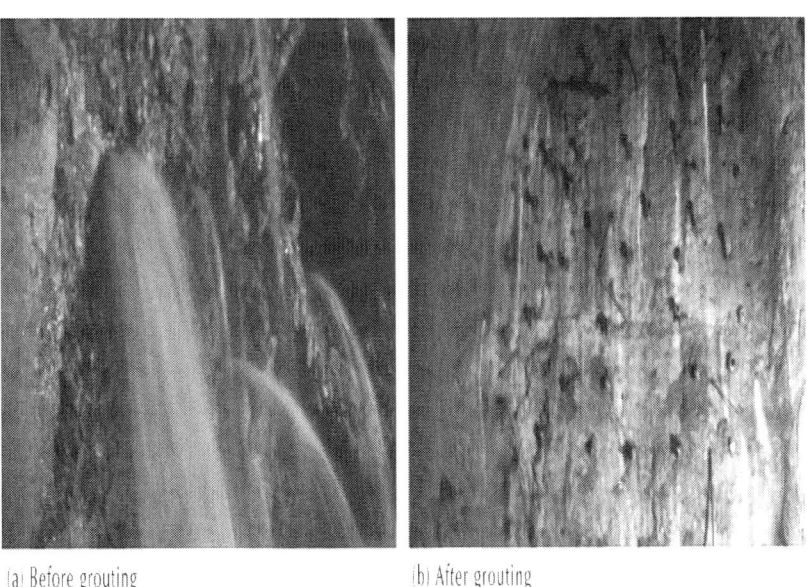

(a) Before grouting   (b) After grouting

Figure 12: Comparison of water inrush in the tunnel face before and after curtain grouting in Guanjiao tunnel.

Application of Advance Level Jet Grouting to Liangshan Tunnel

Liangshan tunnel of Xiamen-Shenzhen Railway is located between Zhangpu Country and Yunxiao County in Fujian Province. The tunnel is 9888 m long, a double-track tunnel with train speed of 250 km/h. During construction, water inrush and mud gushing frequently occurred, resulting in 230 m of main tunnel filled with serious gushed mud. The incident causes a ground subsidence of about 20 m in depth in the area of 1500 m^2(the overburden depth of tunnel is 270 m). The ground subsidence revealed that the surrounding rocks are located in intensively fissured zone.

After the accidents of water inrush and mud gushing, level jet grouting piles were set at the tunnel face in order to (1) ensure the safe construction and long-term operation of the tunnel, (2) to eliminate mud gushing at weak zone in the main tunnel, and (3) to enhance the tunnel stability. Jet grouting piles are 40 m long, over 50 cm in diameter, and 35 cm in spacing between pile centers. The interlocking width of grouting piles is 15 cm and cross-distributed. Four rings of level jet grouting piles are set for the arch wall outside the excavation contour line, and the effective thickness of grouted body is 1.55 m. Two rings of level jet grouting pile are set for the bottom of the inverted arch, and the effective thickness of grouted body is 0.85 m. The compressive strength of grouted body shall be not less than 3 MPa. Advance support of 159 mm large grouting pipe-shed (40 m long) of two layers at arch wall which is inside the inner ring jet grouting piles should be considered. The pipe-shed shall be embedded into the bedrock with depth not less than 5 m, and circular spacing of 0.3 m. After the jet grouting piles are completed, the effect of reinforcement shall be examined and other related reinforcement works shall be carried out before excavation. Thus, the following criteria should be met:

1. The compressive strength of grouted body should be not less than 3 MPa.

2. Water stream is not allowed at testing hole, tunnel face and side wall after excavation.

3. Core sampling rate of jet grouting reinforcement body should be not less than 70%.

If drilling tests results (image borehole tests, core sampling hole and pipe-shed drilling holes) suggest that the jet grouting does not have a targeted effect, additional inspection holes will be drilled in order to determine the potential weak areas; if necessary, supplementary reinforcement measures shall be carried out in the weak areas through supplement pile jet grouting or grouting, or composite grouting to ensure the reinforcement effects of the reinforced rings.

Through the observation in each cycle during excavation, it was observed that the level jet grouting reinforcement ring always keeps its longitudinal continuity in the weak layer. The circular interlocking is good, and the level jet grouting reinforcement rings are visible. Point load test results of level jet grouting reinforced body show that the average strength is 5.7 MPa, with the minimum value of 4.8 MPa; the average compressive strength of the standard grouting samples is 21.7 MPa, with the minimum value of 13 MPa. The values meet the design requirements.

Application of Combined Method to the Baiyun Tunnel

The total length of Baiyun tunnel of Nanning-Guangzhou Railway is 2285 m. The tunnel goes through large regional faults F1, F2. The fault F2 extends more than a dozen of kilometers. The width of fault F2 is 10–40 m, and its deformation influential zone even reaches 100 m. Fault F2 and its secondary fault F1 form a closely paralleled combination of reversed fault, thus the stratum is seriously fragmented and vertical penetration between faults is induced. Within the faults, there are mixtures of breccia and clay of very low intensity. So, the mixtures of breccia and clay pose a very high pressure on the stable layer at the top of the tunnel. The inclination angle of the fault is relatively small, only 25°.

The tunnel construction started from the lower part of the reversed fault. The excavation or blasting unexpectedly disturbed the stratum, and consequently, serious water inrush and mud gushing occurred. As a result, tunnel excavation was stopped as shown in Fig. 13. After the accident of water inrush and mud gushing, many treatment measures were taken (see Fig. 14), including building retaining dam, stabilizing landslide at the tunnel face, draining inrush water, building by-pass heading, jet grouting reinforcement at the top of landslide section in the main tunnel. In addition, curtain grouting was used in the main tunnel. With those measures, the water inrush and mud gushing were successfully stopped.

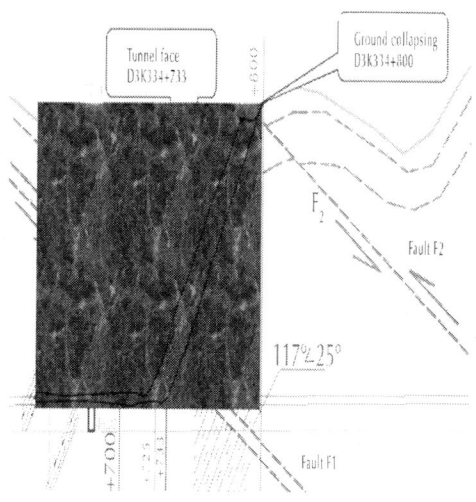

Figure 13: Layout of faults in Baiyun tunnel of Nanning-Guangzhou Railway and the water inrush and mud gushing disaster.

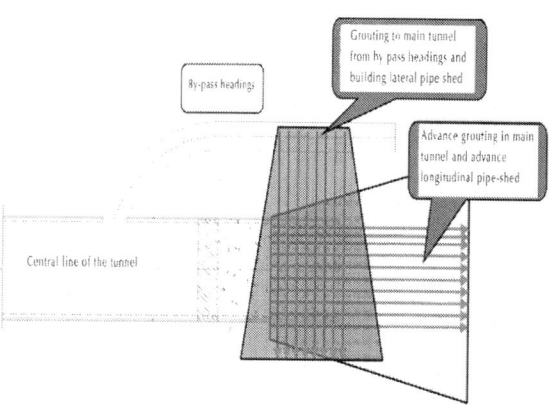

Figure 14: Treatment of water inrush and mud gushing section in Baiyun tunnel of Nanning-Guangzhou Railway.

DISCUSSION

Numerous researches on the karst tunnels with engineering problems such as water inrush are achieved, however, with the rapid development of railway construction in karst region in West China, a series of hot issues concerning safe tunnel construction or operation should be addressed. These issues are listed as follows:

1. Karst water outburst is a dynamic evolution of water inrush induced by long-term water–rock interaction and artificial interferences during tunnel construction. The reported results merely stress the role of karst water in fractured rocks, ignoring the interaction between water and rocks before and after karst water outburst.

2. The accuracy of geological forecast data has an important impact on the construction progress and construction safety of karst tunnels. However, various geological prediction methods currently used have their limitations, such as narrow forecast range, qualitative but not quantitative analysis, inconvenient identification method, and low accuracy. Comprehensive geological prediction methods should be used to improve the accuracy of karst geological prediction. Also, staff training and field identification ability should be improved. We should build up a complete set of applicable karst tunnel geological prediction methods.

3. In the limestone region, fully understanding karst distribution is difficult. Therefore, related geological theory should be focused on to find out the rule of limestone strata distribution in karst regions.

4. Groundwater, especially karst-connection water, is the principal problem to be addressed during karst tunnel construction. At present, water treatment is still considered in two ways by water discharging and blocking, or in combination. However, we should effectively control groundwater discharge, and protect the valuable groundwater resources and the ecological environments.

5. Establishing a system of water prevention and control in karst region is not fully comprehended. We should establish an effective expert decision system in the development regulation

of karst water inrush, the principles of water inrush classification, the model of water inrush, the comprehensive geophysical prospecting technology as well as the early-warning prediction of water inrush.

CONCLUSIONS

1. According to the geological conditions and engineering characteristic of the surrounding rocks, water inrush and mud gushing can be classified into 4 types as shown in Table 1 and Fig. 2.

2. The main factors inducing water inrush and mud gushing are associated with the surrounding rock conditions, meteorological impacts and tunnel excavation. Surrounding rock conditions are the basis of potential mud gushing in karst tunnels. Tunnel excavation is the artificial trigger of water inrush. The meteorological impacts, including temperature and rainfall, will accelerate the progress of water inrush and mud gushing disasters.

3. The macro- and micro-mechanisms of tunnel water inrush and mud gushing are discussed. The essence of tunnel water inrush and mud gushing is the continuous softening and dissolving effect, cavity expansion effect, water wedge effect on the geological interface, and the erosion-expansion effect on water inrush passage that leads to the sudden energy release of water-rich body.

4. Comprehensive advance geological forecast is the principal measure for the prevention of railway tunnel water inrush and mud gushing. Advance geological forecasts are classified into different degrees according to the risks of different sections in the tunnel. The items of forecast shall be carefully determined.

5. The principle of railway tunnel water inrush and mud gushing treatment can be divided into draining- oriented, blocking-oriented or draining-and-blocking. The appropriate method shall be chosen according to the field conditions in practice.

6. Treatment technologies of water inrush and mud gushing include energy relief and pressure reduction technology, advance

grouting technology, and advance jet grouting technology. Each technology has its own benefits and misfits, thus we should consider the suitable technologies based on site-specific geological conditions.

REFERENCES

1. Cui, 2005, J. Cui, Tunnel and underground project construction technology Science Press, Beijing (2005) (in Chinese).
2. Dai, 2009, Y. Dai, Geological features and treatment of mud rushing in Yunwu Mountain tunnel, Railway Engineering, 10 (2009), pp. 33–35 (in Chinese).
3. Gao et al., 1999, Y. Gao, L. Shi, H. Lou, Law of mining floor water-inrush and its preferred plane, Xuzhou: China University of Mining and Technology Press (1999) (in Chinese).
4. Guan and Zhao, 2011, B. Guan, Y. Zhao, Construction technology in tunnels with soft and weak surrounding rocks, China Communications Press, Beijing (2011) (in Chinese).
5. Guan, 2003, B. Guan, Key points in tunnel designing, China Communications Press, Beijing (2003) (in Chinese).
6. Guo, 2011, J. Guo, Study on against-inrush thickness and water burst mechanism of karst tunnel, Ph.D. Thesis Beijing Jiaotong University, Beijing (2011) (in Chinese).
7. Jiang, 2006, J. Jiang, Mechanism and countermeasures of water-bursting in railway tunnel engineering, China Railway Science, 27 (5) (2006), pp. 76–82 (in Chinese).
8. Li, 2010, B. Li, Engineering safety risk analysis and control of subsea tunnel constructed by, drill and blast method, Ph.D. Thesis Beijing Jiaotong University, Beijing (2010) (in Chinese).
9. Li, 2009, L. Li, Study on catastrophe evolution mechanism of karst water inrush and its engineering application of high risk karst tunnel, Ph.D. Thesis Shandong Jiaotong University, Jinan (2009) (in Chinese).
10. Li et al., 2011, S. Li, X. Zhang, Q. Zhang, Research on mechanism of grout diffusion of, dynamic grouting and plugging method in water inrush of underground engineering, Chinese Journal of

Rock Mechanics and Engineering, 30 (12) (2011), pp. 2377–2396 (in Chinese).

11.   Lin and Song, 2012, G. Lin, R. Song, Research on the mechanism and treatment technology of mud gushing in karst tunnel, Tunnel Construction, 32 (2) (2012), pp. 169–174 (in Chinese).

12.   Mo and Zhou, 2008, Y. Mo, X. Zhou, Dynamic monitoring and simulation analysis of surrounding rock deformation of tunnel in karst region, Chinese Journal of Rock Mechanics and Engineering, 27 (Suppl. 2) (2008), pp. 3816–3820 (in Chinese).

13.   Yang and Zhang, 2008, D. Yang, P. Zhang, Study on the rapid treatment method of mud gushing and landslide at shallowly covered karst section in Yunwu Mountain tunnel, West-China Exploration Engineering (10) (2008), pp. 192–194 (in Chinese).

14.   Zhang, 2010, M. Zhang, Tunnel construction technology in karst fault of Yiwan Railway, Science Press, Beijing (2010) (in Chinese).

15.   Zhang et al., 2010, X. Zhang, S. Li, Q. Zhang, H. Li, W. Wu, R. Liu, et al., Filed test of comprehensive treatment for high pressure dynamic grouting, Journal of China Coal Society, 35 (8) (2010), pp. 1314–1318 (in Chinese).

16.   Zhao et al., 2009, Y. Zhao, S. Tian, Z. Cao, Geological work method for the construction of the Yichang-Wanzhou Railway tunnel in high-risk karst areas, Journal of Shandong University (Engineering Science) (5) (2009), pp. 91–95 (in Chinese).

17.   Zhao, 2012, Y. Zhao, Study on deformation mechanism and control technology of weak rock surrounding tunnel, Ph.D. Thesis Beijing Jiaotong University, Beijing (2012) (in Chinese).

18.   Zhuang and Mu, 2009, H. Zhuang, J. Mu, The prevention and treatment of large cross section tunnel mud gushing in karst areas, Railway Engineering (6) (2009), pp. 49–51 (in Chinese).

Citations

CHAPTER 1

Rachel A. Grant, Tim Halliday, Werner P. Balderer, Fanny Leuenberger, Michelle Newcomer, Gary Cyr, and Friedemann T. Freund, Ground Water Chemistry Changes before Major Earthquakes and Possible Effects on Animals, doi: 10.3390/ijerph8061936.

CHAPTER 2

C. Mullen (2013). FE Based Vulnerability Assessment of Highway Bridges Exposed to Moderate Seismic Hazard, Engineering Seismology, Geotechnical and Structural Earthquake Engineering, Dr Sebastiano D'Amico (Ed.), ISBN: 978-953-51-1038-5, InTech, DOI: 10.5772/55334.

CHAPTER 3

Liber Galban (2010). Model for Geological Risk Management in the Building and Infrastructure Processes, Advances in Risk Management, Giancarlo Nota (Ed.), ISBN: 978-953-307-138-1,

CHAPTER 4

Robin Bronen and F. Stuart Chapin, Adaptive Governance and Institutional Strategies for Climate-Induced Community Relocations in Alaska, doi: 10.1073/pnas.1210508110.

CHAPTER 5

Shannon Doocy Amy Daniels, Sarah Murray, and Thomas D. Kirsch, The Human Impact of Floods: a Historical Review of Events 1980-2009 and Systematic Literature Review , doi: 10.1371/currents.dis.f4d eb457904936b07c09daa98ee8171a.

CHAPTER 6

John T. Watson, Michelle Gayer, and Maire A. Connolly, Epidemics after Natural Disasters, doi: 10.3201/eid1301.060779.

CHAPTER 7

Ourania Lasda, Angela Dikou, and vangelos Papapanagiotou, Flash Flooding in Attika, Greece: Climatic Change or Urbanization, doi:10.1007/s13280-010-0050-3.

CHAPTER 8

Pierre Duffaut, The traps Behind the Failure of Malpasset Arch Dam, France, in 1959, doi:10.1016/j.jrmge.2013.07.004.

CHAPTER 9

Shengwu Song, Xuemin Feng, Hongling Rao, and Hanhuai Zheng, Treatment Design of Geological Defects in Dam Foundation of Jinping I Hydropower Station doi:10.1016/j.jrmge.2013.06.002.

CHAPTER 10

D.A. Gunn, J.E. Chambers, S. Uhlemann, P.B. Wilkinson, P.I. Meldrum, T.A. Dijkstra, E. Haslam, M. Kirkham, J. Wragg, S. Holyoake, P.N. Hughes, R. Hen-Jones, and S, Glendinning, Moisture Monitoring in Clay Embankments Using Electrical Resistivity Tomography, doi:10.1016/j.conbuildmat.2014.06.007.

CHAPTER 11

Yong Zhao, Pengfei Li, and Siming Tian, Prevention and Treatment Technologies of Railway Tunnel Water Inrush and Mud Gushing in China, doi:10.1016/j.jrmge.2013.07.009.

Index